PLANT ANATOMY

PLANT ANATOMY

FORREST F. STEVENSON
THOMAS R. MERTENS
Ball State University
Muncie, Indiana

John Wiley & Sons, Inc.
New York • London • Sydney • Toronto

Editors: Judy Wilson and Irene Brownstone
Production Manager: Ken Burke
Editorial Supervisor: Winn Kalmon
Composition and Make-up: Winn Kalmon
Artist: Design Services, Muncie, Indiana

Library of Congress Cataloging in Publication Data

Stevenson, Forrest Frederick, 1916-
Plant anatomy.

(Wiley self-teaching guides)
Bibliography: p.
Includes index.
1. Botany—Anatomy—Programmed instruction.
I. Mertens, Thomas Robert, 1930- joint author.
II. Title.
QK641.S73 582'.04'077 75-23285
ISBN 0-471-82460-7

To the Reader

Today, perhaps as never before, we are becoming increasingly concerned with our supply of natural resources, the relationships between ourselves and the environment, and how pollution of our world may alter these relationships. In short, we have become ecologically minded. Possibly as a result of a new awareness of living things around us, many persons are "discovering" plants for the first time as they learn of the critical role which plants play in the ecological system. They are discovering, too, what many have long known— that cultivating plants is a pleasant and relaxing hobby.

Whether we deal with them as biologists, environmentalists, or hobbyists, we can benefit from an anatomical study of plants, for the functions of a plant are directly related to its structure. When we possess a knowledge of the internal structure of plants, therefore, we may be better equipped to evaluate the extravagant claims about plant behavior often advanced by nonbotanists.

There are over 300,000 known kinds of plants in the world. Many of them, such as algae, fungi, and mosses, possess a simple organization. While these simple plants are interesting in their own right, this book deals solely with the flowering plants which are the most common and best known around us. Flowering plants are of great economic importance, providing us with food, clothing, and various commercial products.

The presentation of plant anatomy in this book begins with a discussion of the characteristics of plant cells and progresses through a study of primary and secondary tissues. These tissues, in turn, are identified as components of the organs of the plant, such as roots, stems, leaves, and flower parts. The functions of the tissues and plant organs are also considered.

A book of this nature and size must, of necessity, deal in generalities, and the authors are aware that there are exceptions to certain statements made in the book. Nevertheless, they believe that the teaching of primary concepts should precede the presentation of exceptional conditions. For its intended audience this book's presentation will be a more than adequate introduction to the anatomy of flowering plants.

February 1976

F. F. S.
T. R. M.

How to Use This Book

This book is designed so that you can learn about the anatomy of flowering plants enjoyably and thoroughly at your own pace. The material is presented in sections called frames, each containing questions and answers that let you test your understanding of the new material. As you come to a question, pause and answer in the space provided. Your answer may be a true-false or multiple-choice selection, a word or short phrase to fill in a blank, an explanation, or labels on a simple diagram; it may involve either reasoning or remembering. Check your answer with the answer given following the dashes that separate each question from its answer. For maximum learning, try not to look at the printed answer until you have first given your own. (You may want to cover the answers with a separate sheet of paper or a 3 x 5 card while you are working on the question.)

You will be able to answer practically all of the questions as you come to them because the questions progress from simple to complex and each frame builds upon information in the previous frames. This learning program is <u>not</u> a workbook and it is <u>not</u> a textbook. Unlike a workbook this program is self-contained and provides answers for all the questions it asks. Unlike a textbook the program helps you to learn and, through frequent problems and review items, helps you to be confident that you can use the material it presents.

Work at your own pace. You may be able to move rapidly through some sections but only slowly and thoughtfully through others. Some learners may complete the program in four hours, others in sixteen, a typical figure probably being eight or ten hours—probably less time than you would spend learning the same material from textbook reading assignments, class lectures and discussions, and studying for a test.

You may wish to do only one chapter at a time, with a break (even a day) in between. Because of the continual review in answering questions, you will not forget one chapter if you wait awhile before beginning the next.

Review items occur throughout the program, and each chapter concludes with a Self-Test. If you are in a biology or botany course, you will find all of these sections helpful in preparing for a classroom test. A Final Test is provided at the end of the book, so that you can evaluate your overall learning.

Contents

Behavioral Objectives and Field Testing

To assist you in learning the content of this book, we have prepared a precise set of behavioral or instructional objectives. These objectives indicate the kinds of behaviors that you should be able to demonstrate once the content of the book has been mastered. Upon completion of this book you will be able to do the following.

- Characterize plant cells by describing the relationship between the protoplast and the cell wall.
- Describe the relationship between protoplasm, cells, tissues, and organs.
- Characterize by location and function apical, intercalary, and lateral plant meristems.
- Compare and contrast the five simple plant tissues with respect to structure and function.
- Compare and contrast simple versus complex plant tissues.
- Compare and contrast xylem and phloem with respect to structure and function.
- Identify by description and function and in drawings or photographs the principal cell types found in xylem and phloem.
- Outline the development of the embryo of a flowering plant, indicating the origin and function of the different parts of the embryo.
- Describe the origin of various seedling tissues and organs from the parts of the embryo.
- Differentiate between monocots and dicots and give examples of each.
- List four general functions of roots.
- List three ways in which roots may originate.
- Identify and describe the two major kinds of root systems and cite examples and advantages to the plant of each.
- Compare and contrast the anatomy of monocot and dicot roots.

- Cite the functions of such root parts as the epidermis, cortex, endodermis, pericycle, and stele.
- Identify, compare, and contrast the four major regions discernible along the length of a growing root.
- List at least three general functions of stems.
- Compare and contrast the anatomy of monocot and dicot stems.
- Describe the external features of a woody stem by labeling those features on an appropriate diagram.
- Compare and contrast roots and stems with respect to both internal and external anatomy, identifying those features which are characteristic of each as well as those they have in common.
- Identify and give specific examples of specialized stems such as corms, bulbs, rhizomes, and tubers that function in food storage.
- Describe the origin of the vascular and cork cambiums in perennial dicot roots and stems.
- Identify secondary xylem, secondary phloem, and the vascular cambium in a diagram or photograph of a cross section of a perennial dicot root or stem.
- Describe the fate of the primary plant tissues (primary xylem, primary phloem, pericycle, endodermis, cortex, and epidermis) as secondary growth takes place.
- Identify three functions of pericycle in a perennial dicot root.
- Describe the origin and development of annual growth rings, springwood, and summerwood in a perennial dicot stem.
- Differentiate between heartwood and sapwood in terms of location, function, and color.
- Describe the appearance, cellular constituents, location, and function of vascular rays.
- Compare and contrast information gained about wood (secondary xylem) anatomy from studying cross, tangential, and radial sections.
- Describe the external and internal structure of a typical dicot leaf, identifying various parts revealed in a magnified leaf cross section.
- Compare and contrast the leaves of monocot and dicot plants with respect to venation pattern, presence or absence of a petiole, and internal structure.
- Relate various leaf structures (such as veins, stomates, mesophyll, and chloroplasts) to the photosynthetic function of the leaf.
- Identify a leaf from a drawing or photograph as monocot or dicot, simple or pinnately or palmately compound, pinnately or palmately veined.

- Identify flower parts in a photograph or diagram of a flower, indicating whether the parts are essential or nonessential to seed formation.
- Identify the component parts and the function of the essential flower parts.
- Differentiate between a simple and compound pistil and identify in diagrams the placentation types that may occur in compound pistils.
- Define fruit in terms of its origin and differentiate between various types of fruits, such as fleshy (berry, drupe, pome), dry dehiscent, and dry indehiscent (achene, nut, grain).
- Use the following vocabulary of plant anatomy to describe the internal and external structure of a vascular plant.

Chapter 1

cambium
cell
cell wall
    primary
    secondary
cellulose
collenchyma
cork
cuticle
cutin
epidermis
fiber
guard cell
lignin
meristems
    apical
    intercalary
    lateral
middle lamella
organ
parenchyma
pectin
pit
pit membrane
pit pair
pasmodesma (plural, plasmodesmata)
protoplasm
protoplast
sclereid
sclerenchyma
stoma (plural, stomata)
stone cell
suberin
tissue
    meristematic
    permanent
    primary
    secondary
    simple

Chapter 2

aperture
companion cell
complex tissue
fiber tracheid
libriform fiber
lumen
nucleus (plural, nuclei)
phloem
phloem fiber
phloem parenchyma
pit
    bordered
    simple
pit cavity
sieve area
sieve plate
sieve tube
sieve tube member
thickening
    annular
    reticulate
    scalariform
    spiral
torus
tracheid
vascular tissue

Chapter 2 (continued)

vessel
vessel member
xylem
xylem fiber
xylem parenchyma

Chapter 3

cotyledon
dicotyledon (dicot)
egg cell
embryo
epiblast
epicotyl
germinate
ground meristem
ground tissues
hypocotyl
monocotyledon (monocot)
procambium
proembryo
protoderm
radicle
seedling
sperm cell
suspensor

Chapter 4

adventitious (root)
Casparian strip
cortex
endodermis
endogenous
epidermis
fibrous root system
pericycle
pith
primary root
region of differentiation
region of elongation
rootcap
root hair
root hair region
secondary root
stele
taproot system
tertiary root

Chapter 5

annual
bud scale
bulb
chlorenchyma
corm
exogenous
herbaceous
internode
lateral bud
leaf primordia (singular, primordium)
leaf scar
node
perennial
pith ray
rhizome
terminal bud
tuber
vascular bundle
- bicollateral bundle
- collateral bundle

vascular cambium

Chapter 6

annual growth rings
axial parenchyma
cork cambium
heartwood
periderm
phelloderm
phellogen
phellum
radial section
sapwood
springwood
summerwood

tangential section
vascular ray
- phloem ray
- xylem (wood) ray

wood (secondary xylem)

Chapter 7

abscission layer
anther
blade
bulliform cell
bundle sheath
carpel
chlorophyll
chloroplast
chromoplast
compound leaf
- palmately compound
- pinnately compound

dorsal bundle (of a carpel)
filament
flower parts
- essential
- nonessential

fruit
- dry (dehiscent and indehiscent)
- fleshy

growth
leaflet
locule
mesophyll
- palisade
- spongy

ovary
ovule
pericarp
- endocarp
- exocarp
- mesocarp

petal
petiole
photosynthesis
pistil
- compound
- simple

placenta
placentation
- axile
- free-central
- parietal

pollen grain
protective layer
rachis
receptacle
sepal
sheath
simple leaf
stamen
stigma
stipules
style
veins
venation
- net
- palmate
- parallel
- pinnate

ventral bundle (of a carpel)
whorls

Plant Anatomy was field-tested with 65 beginning plant science students in the Department of Biology at Ball State University, Muncie, Indiana. During the first week of instruction in the spring quarter, 1975, each of the students was given a 150-item pretest. The mean score on this test was determined to be 49.49 points or 33 percent. After a two-week interval for study of Plant Anatomy, post-testing began, using a post-test identical to the pretest.

The post-test was divided into three parts and given at one week intervals. The first post-test dealt with the content of the first two chapters in Plant Anatomy, while the second post-test dealt with Chapters 3, 4, and 5. The final post-test treated Chapters 6 and 7. The total possible score on the post-test was 150 points and the mean score was calculated to be 128.32 points or 85.5 percent.

A gain score was calculated using the following formula recommended by Professor McGuigan of Hollins College, Virginia.

$$\% \text{ Gain} = \frac{\text{Post-Test Score Minus Pretest Score}}{100\% \text{ Minus Pretest Score}}$$

Thus, using the scores indicated above,

$$\% \text{ Gain} = \frac{128.3 - 49.5}{150 - 49.5} \times 100 = \frac{78.8}{100.5} \times 100 = 78\%.$$

The pre/post-test was modified to form the final test in this book. Plant Anatomy was revised according to students' testing experience, as well as subject matter and programming reviewers' comments, with the goal of further improving its teaching effectiveness.

The authors and publisher are interested in data on the use of this book in other settings. Please send any pre- and post-test scores on Plant Anatomy, along with other comments or suggestions, to the Self-Teaching Guides Editor, John Wiley & Sons, 605 Third Avenue, New York, New York 10016.

Reference Chart for Selected Biology and Botany Textbooks

You may find the information in the table on page xix useful if you wish to pursue certain topics in greater depth or if you are a student in a course using one of the books listed below. The table correlates chapters in this book with material on the same subjects in a number of the newer basic biology and botany textbooks. The numbers listed in the table correspond to chapter (and part) numbers in the textbooks.

Bold, H. C., Morphology of Plants, 3rd ed. (New York: Harper and Row, 1973).

CRM Books Editorial Staff, Biology Today, 2nd ed. (Del Mar, California: CRM Books, 1975).

Cronquist, A., Basic Botany, 3rd ed. (New York: Harper and Row, 1973).

Curtis, H., Biology, 2nd ed. (New York: Worth, 1975).

Greulach, V. A. and J. E. Adams, Plants: An Introduction to Modern Botany, 3rd ed. (New York: Wiley, 1976).

Haynes, J. D., Botany—An Introductory Survey of the Plant Kingdom (New York: Wiley, 1975).

Jensen, W. A. and F. B. Salisbury, Botany: An Ecological Approach Belmont, California: Wadsworth, 1972).

Keeton, W. T., Biological Science, 3rd ed. (New York: Norton, 1972).

Keeton, W. T., Elements of Biological Science, 2nd ed. (New York: Norton, 1973).

Muller, W. H., Botany: A Functional Approach, 2nd ed. (New York: Macmillan, 1969).

Neushul, M., Botany (Santa Barbara, California: Hamilton, 1974).

Raven, P. H. and H. Curtis, Biology of Plants (New York: Worth, 1970).

Stephens, G. C. and B. B. North, Biology (New York: Wiley, 1974).

Weier, T. E., C. R. Stocking, and M. G. Barbour, Botany—An Introduction to Plant Biology, 5th ed. (New York: Wiley, 1974).

Wilson, C. L., W. E. Loomis, and T. A. Steeves, Botany, 5th ed. (New York: Holt, Rinehart and Winston, 1971).

Chapter in this book	Bold	CRM Books Staff	Cronquist	Curtis	Greulach & Adams	Haynes	Jensen & Salisbury	Keeton (Science)	Keeton (Elements)	Muller	Neushul	Raven & Curtis	Stephens & North	Weier, Stocking, & Barbour	Wilson, Loomis, & Steeves
1. Plant Cells and Simple Tissues	17	13	4, 18	6	4	20, 26	24	3	3	3, 5	13	1(1), 1(3), 2(1)	3	4, 7	3, 7
2. Complex Tissues: Xylem and Phloem	17	19	18	24	4	20, 26	24	3, 7	3, 7	6, 8	13	2(1)	7	7	7
3. Embryos and Seedlings	30	19	25	26	5, 10	26	24	16	16	26	17	2(2)	13	3, 16	12, 13, 14
4. The Root	17, 30	19	20	24, 26	5, 20	20, 26	24	5	5	8	13	2(2)	7	9	2, 7
5. The Stem	17, 30	19	19	24, 26	5, 20	20, 26	24	6, 7, 16	6, 7, 16	6	13	2(2)	7	7	2, 7
6. Secondary Growth	17	19	19, 20	26	5	20, 26	24	16	16	6, 8	13	2(2)	7	7, 8, 9	7, 9
7. Leaves, Flowers, and Fruits	17, 30	5, 19	21, 25	21, 24, 26	5	20, 26	24, 28	4, 6, 21	4, 6, 21	9, 26	13, 17	2(2), 6(7), 6(8)	7, 12	10, 15, 16	2, 5, 12, 13

CHAPTER ONE
Plant Cells and Simple Plant Tissues

1. Few of us have difficulty in deciding whether an object is living or nonliving, but we may be hard pressed to explain our decision. Even biologists may be frustrated in listing clear cut distinctions between living and nonliving things. The many characteristics exhibited by a living organism are the properties of protoplasm—that unique substance of which all living things are composed. Protoplasm is so closely associated with life that a famous biologist described it as "the physical basis of life."

   With few exceptions, protoplasm does not function in large undivided masses. A microscopic examination of a living organism reveals that protoplasm is divided into tiny structural units which are called cells.

   Animals and plants constitute our living world. What two characteristics do animals and plants have in common? ____________________

   ____

   - - - - - - - - - - - - - - - - -

   They are both composed of protoplasm; they are both composed of cells.

2. Because living material is composed of cellular units analogous to the bricks which form the units of a building, and because all protoplasmic activity occurs within each unit, the cell may be described as a "unit of structure and function."

   Considering the information presented in this frame, could you logically expect to find single individual cells existing independently of other cells? Explain your answer. ____________________

   ____

   ____

   - - - - - - - - - - - - - - - - -

   Yes, since the cell is the basic unit of structure and function. Many such organisms, in fact, do exist in nature. However, in multicellular plants and animals, the individual cells function in coordinated groups.

3. You may not experience difficulty in distinguishing living organisms from nonliving objects, but can you distinguish between the many kinds of living organisms? Can you list clear distinctions which separate plants from animals? The task is not easy, and the ready replies that animals move and plants do not, or that plants are green and animals are not, lose their validity with careful consideration.

When you recall that the activities of the organism occur at the cellular level, you may expect that the primary distinction between animals and plants is made at this same level.

A plant cell, unlike an animal cell, secretes a wall of nonliving material around itself during its development. Thus, the protoplasm of the plant cell is enclosed by a cell wall. Each plant cell stands out clearly from its neighbors when viewed under the microscope because of its confining and rigid cell wall (see photograph).

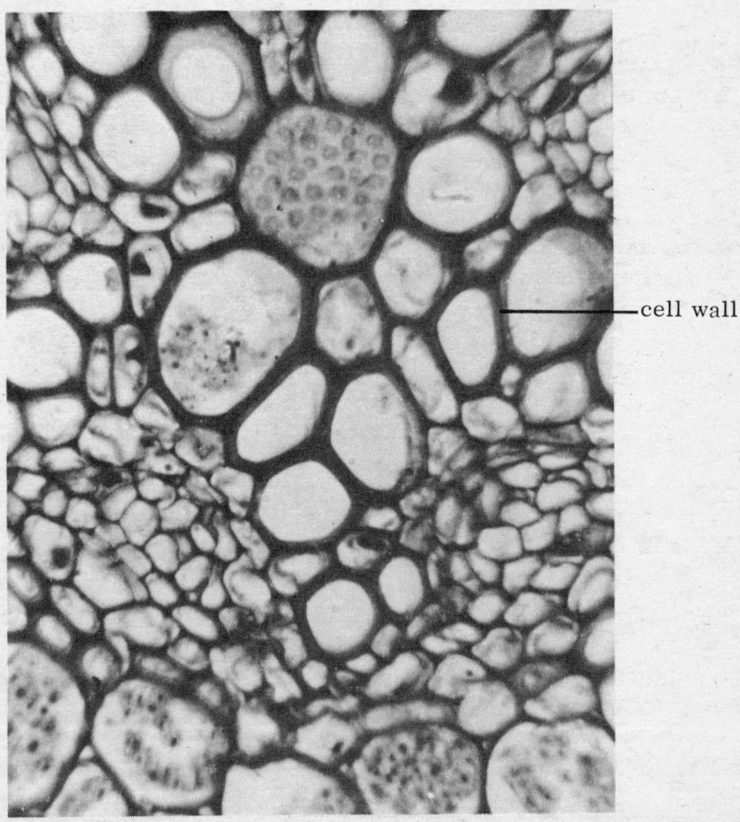

If you scraped some dead cells from the inside lining of your cheek and examined them under the microscope, would you expect the individual cells to stand out clearly? Explain. ______________________

____

- - - - - - - - - - - - - - - - - -

No, because these individual animal (human) cells are not enclosed by cell walls.

4. Plainly, the plant cell is composed of two main parts: (1) the cell wall, and (2) the protoplasm which is given the convenient name protoplast to distinguish the living portion of the cell from the cell wall. A protoplast may be defined as that portion of a plant cell exclusive of the cell wall.

Although the word protoplast may be applied to both plant and animal cells, why does it have special significance in relation to plant cells?

____

____

- - - - - - - - - - - - - - - - -

The term protoplast clearly refers to the living material or protoplasm of a plant cell in contrast to the nonliving cell walls. Thus, the term protoplast allows us to distinguish easily between the two essential parts of a plant cell, the living protoplasm and the nonliving cell walls.

5. Cell walls of plants vary considerably in composition and appearance from one cell type to another. They often determine the role a cell may play in the organ of a plant even though the walls are nonliving secretions of the protoplasm and not composed of protoplasm themselves.

The materials which form the cell wall are secreted by the protoplast, while the cell is still increasing in size. The newly formed cell, therefore, is surrounded by a cell wall by the time its growth has been completed. The wall thus formed by the growing cell may vary in thickness and is called the primary cell wall. Primary cell walls are composed largely of the carbohydrates cellulose and pectin.

Check the true statement(s) among the following.

________ a. The cell wall is a living part of the plant cell.

________ b. The plant cell wall is secreted by the protoplast of the plant cell.

________ c. The term protoplast includes both the plant cell wall and the living part of the plant cell.

________ d. The wall formed around a growing plant cell is called the primary cell wall.

- - - - - - - - - - - - - - - - -

b and d are correct

6. You may wonder at this point how the many individual cells which comprise a plant are held together. A thin layer of cementing material lies

between the primary cell walls of two adjacent cells. This layer is called the middle lamella and is responsible for holding the cells of a plant together (see diagram).

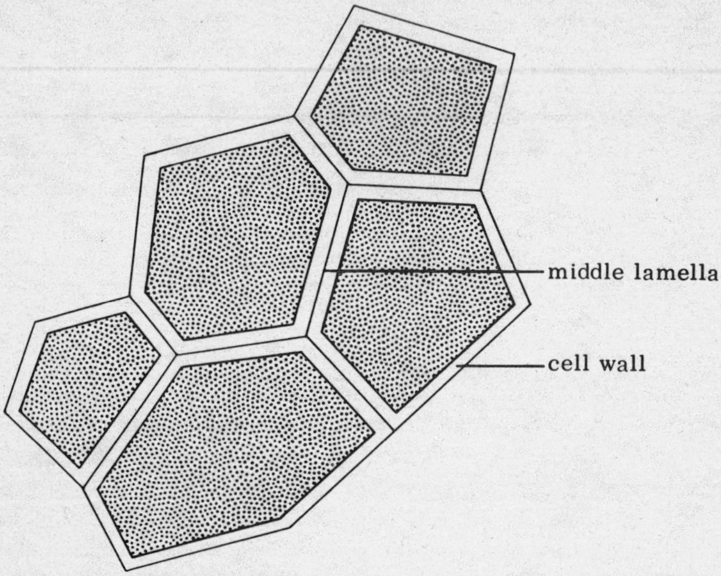

Pieces of plant material when soaked in certain strong acids disintegrate into the separate individual cells which comprised them. The plant material is then said to be macerated. Suggest how the acids may have accomplished maceration. ______________________________

____

- - - - - - - - - - - - - - - - -

The acids apparently dissolved the middle lamella between each of the cells, thus allowing the cells to separate from one another.

7. Although the cells of a plant appear to be completely isolated from each other by their cell walls, such is not the case, for the cell walls are porous enough to allow water and dissolved materials to flow freely from one cell to another. Moreover, the primary cell walls are not always laid down in a continuous and uninterrupted fashion. Thin circular areas called pits are thus formed by unequal deposition of wall material.

Generally, a pit formed in the primary wall of one cell is matched by a similar one formed in the same area of an adjoining cell. Thus, two pits (pit pairs) are separated from each other by the middle lamella and the two thin primary walls which together form a pit membrane (see diagram on next page).

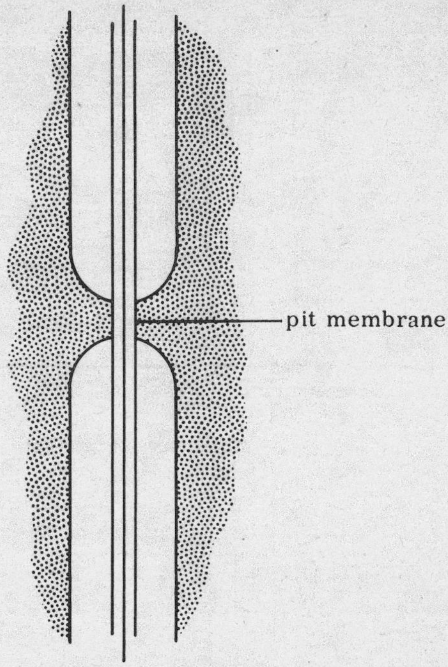

Pit pairs are characterized by tiny strands of living protoplasm which extend through pores in the thin membrane separating them. Therefore, the protoplasts of adjoining cells are linked by living bridges called plasmodesmata (singular, plasmodesma). While the true significance of plasmodesmata is a matter of speculation, it is intriguing to note that the protoplasts of most, if not all, living cells are physically joined to each other by a network of protoplasm. Thus, cells do not appear to be isolated entities.

The diagram on the next page illustrates several plasmodesmata linking two protoplasts. Label the parts of the pit membrane to illustrate its compound nature. (The answers are given on this page, following the dashed line, so cover them before you look at the diagram and then refer back to check your answers.)

- - - - - - - - - - - - - - - - -

a. primary wall; b. middle lamella; c. primary wall

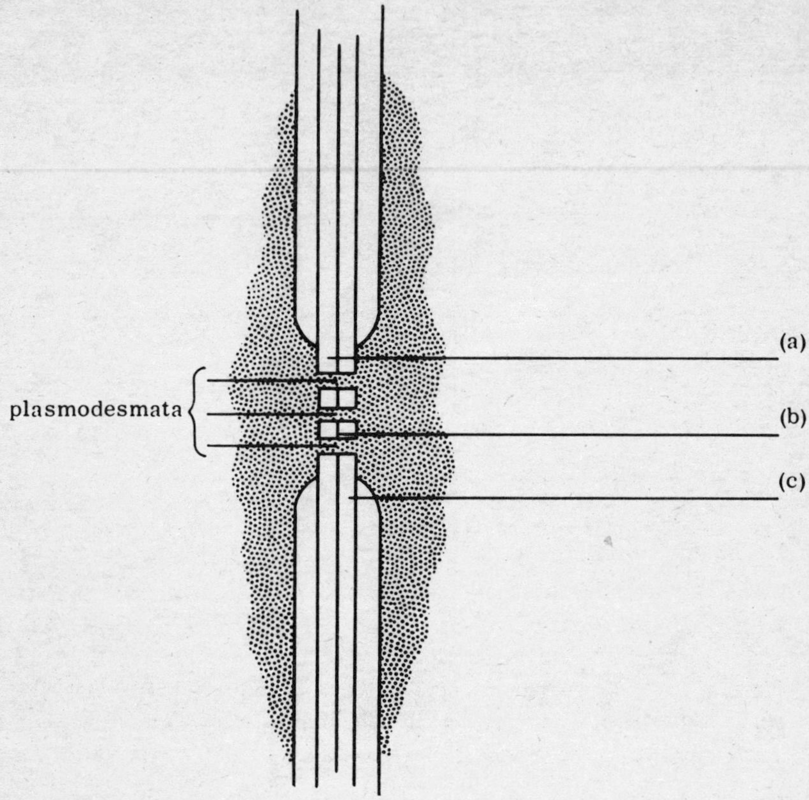

8. Some mature plant cells function while enclosed solely by their primary cell walls. The protoplasts of many other cells secrete additional wall material between the primary wall and the outer boundary of the protoplast itself. This layer formed after the cell has ceased to grow becomes the <u>secondary cell wall</u> and adds additional thickness to the cell wall (see diagram on next page).

Check the statement(s) which are correct.

________ a. The secondary wall lies closer to the protoplasm than the primary wall.

________ b. A secondary wall may be formed before the cell finishes its growth.

________ c. All plant cells form secondary cell walls.

________ d. The primary cell wall is completely replaced by the secondary cell wall.

- - - - - - - - - - - - - - - - - -

a is correct

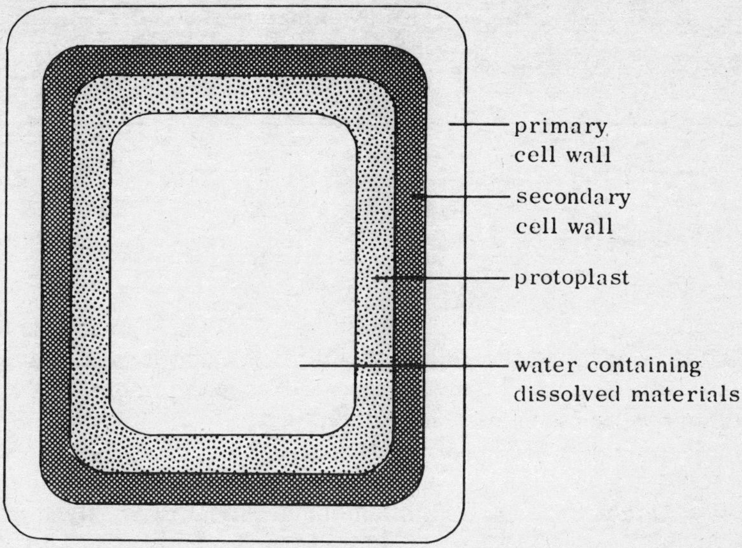

9. Like the primary cell wall, the secondary wall, when present, may vary in thickness. Moreover, the secondary wall may be deposited unevenly on the inside of the primary cell wall so that some parts of the wall are thicker than others.

Shown below are a diagram and a photograph illustrating cells with spiral thickenings.

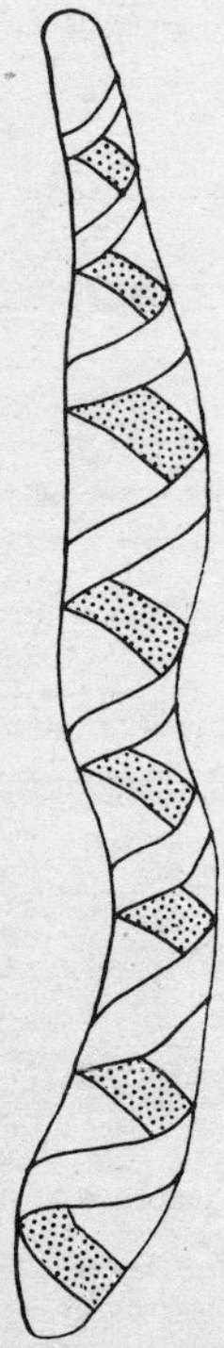

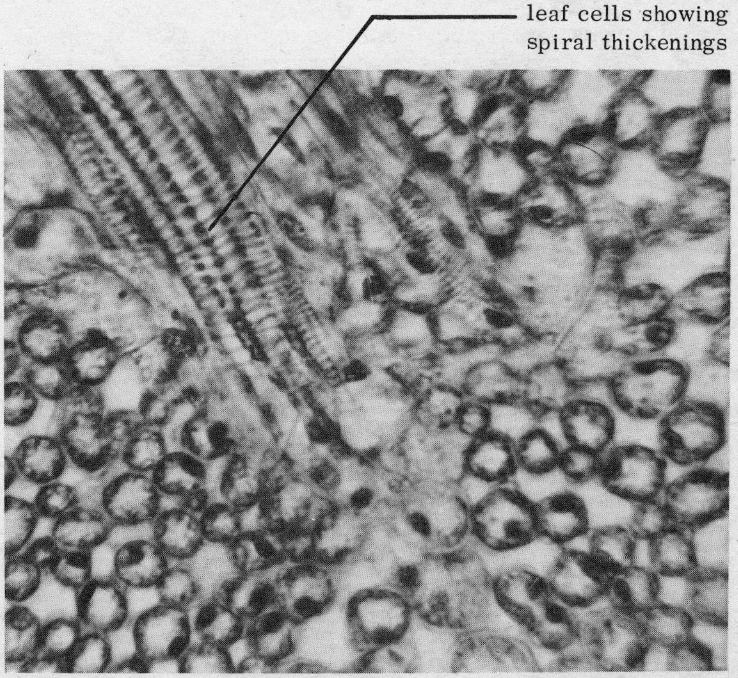

Can you guess how the spiral thickenings of the cell walls in the two illustrations were formed? ______________________________

____

- - - - - - - - - - - - - - - -

The secondary cell wall has been deposited on the primary cell wall in a spiral manner.

10. Up to this point the description of a plant cell has emphasized the structural appearance of the cell wall with little mention of the living protoplast, but the activities of the living cell are of utmost importance in determining the nature of the cell.

The individual cells of any living multicellular organism do not function independently of each other, but instead they are organized into groups which perform certain functions which are required to keep the organism alive. Such a group of cells performing a specific function is called a <u>tissue</u>. The physiological or functional differences between various kinds of tissues are often accompanied by differences in physical appearances as well, since the cells of the tissues are adapted to the tasks they perform.

a. Individual cells organized into groups to perform a specific function are called ____________________.

b. If a group of cells is organized to perform a specific function, would you expect these cells to have a different appearance from cells performing another function? Justify your answer. ______________

____

- - - - - - - - - - - - - - - -

a. tissues
b. Yes, because the appearance of each kind of tissue reflects its adaptations to a specific physiological activity.

11. When examining a flowering plant a plant anatomist encounters many tissues which comprise the plant body. He finds that the familiar parts of a plant such as roots, stems, and leaves differ from each other primarily in the manner in which the tissues are assembled. Such an assemblage of tissues working together is called an <u>organ</u>.

Animals, too, are composed of such tissues as connective, vascular, muscular, nervous, and epithelial. Human anatomists find that the stomach is composed of all of these tissues.

Can the human stomach be considered an organ? Explain your answer.

____

- - - - - - - - - - - - - - - -

Yes, the stomach is an organ because it is composed of several tissues.

12. To review the relationship of parts and to summarize the organization that occurs in plants, place the following words in sequence from simplest to most complex.

(1) cells, (2) organs, (3) protoplasm, (4) tissues

____

- - - - - - - - - - - - - - - - -

(3) protoplasm, (1) cells, (4) tissues, (2) organs

13. Now, write a brief paragraph summarizing your understanding of the relationship between protoplasm, cells, tissues, and organs.

____

____

____

____

____

____

- - - - - - - - - - - - - - - - -

Your paragraph should include the following concepts: Cells are the structural and functional units constituting living things. Cells are made up of the living substance, protoplasm. Groups of cells organized to perform specific functions are called tissues. An assemblage of different tissues working together constitute an organ.

14. Because tissues play an important role in the structure of plants, it is important to learn something about them. For convenience, we can divide plant tissues into two large categories: (1) meristematic tissues whose cells divide and enlarge, and (2) permanent tissues whose cells no longer divide and grow.

If you cut off the tip of a blade of grass, the top of the remaining grass blade does not change in appearance during the rest of the summer. The blade of grass, however, continues to elongate due to new growth at the base of the leaf where it joins the stem.

From this you may conclude that the tissues at the leaf base are

______________________ and the tissues of the leaf closer

to the tip are ______________________.

- - - - - - - - - - - - - - - - -

meristematic; permanent

15. Young animals experience a general growth of all tissues as they increase in size, and growth is not completely restricted to certain specific body regions. In contrast, a plant possesses special restricted areas of growth which are established at the time of early embryo development. These growth regions often persist throughout the life of the plant and constitute the <u>meristems</u>.

Can the term <u>meristem</u> be used to describe growth of the human child? Explain. ______________________________

____

- - - - - - - - - - - - - - - - -

No, since growth is not restricted to limited areas or regions in the body of a growing child.

16. Meristems do not occur at random in a plant but are confined to certain predictable areas. For that reason, it is possible to group meristems according to location. The general organization of a flowering plant is simple in spite of the apparent complexity which the branching of the roots and stems conveys.

A plant may be visualized as an axis divided into the root and the stem to which the leaves are attached as appendages.

Growth of the plant axis occurs at the apices or tips of the stem and root. Thus, these meristems may be classified as <u>apical</u> meristems (see diagram on next page).

A nail is driven into the trunk of a tree three feet above the ground and the nail remains in the tree for ten years. Neglecting for now the obvious increase in diameter of the stem, indicate by a check mark which of the following events you would expect to observe at the end of the ten years.

________ a. The nail will be higher than three feet from the ground.

________ b. The nail will remain three feet above the ground.

________ c. The nail will be lower than three feet above the ground.

- - - - - - - - - - - - - - - - -

b is correct. The nail will remain at the original distance from the ground because all upward growth of the stem has occurred at the apical meristem.

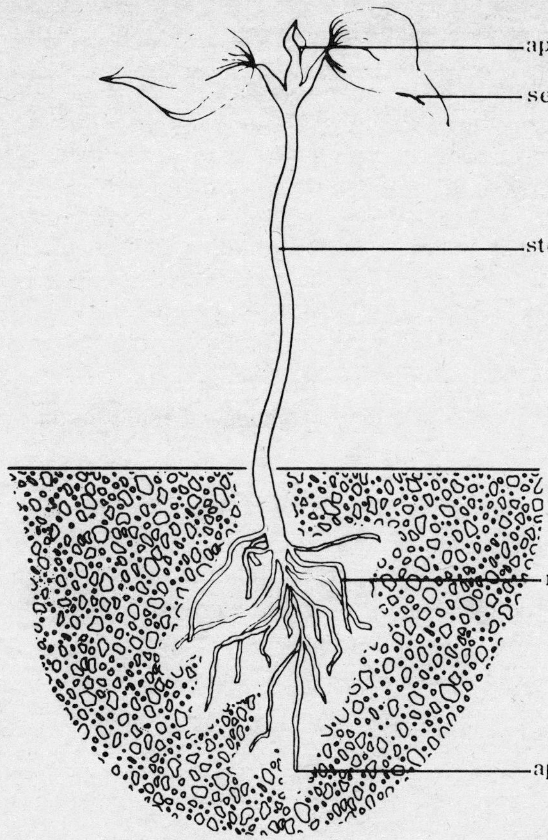

17. Growth of the stem tips and the root tips increases the length of the axis. Obviously extensive linear growth cannot continue indefinitely without provision for an increase in the diameter of the axis as well. A ring or cylinder of meristematic tissue develops near the perimeter of the axis of many plants to form the lateral meristem, from which additional tissues arise to add to the girth of the axis.

The name cambium is applied to the lateral meristem and it is used almost exclusively to refer to any kind of lateral meristem.

Bottle cork is obtained by stripping the outer bark from the trunks of certain oak trees. A single tree may yield several layers of cork through repeated strippings over the years.

What type of meristem produces cork? ______________________

- - - - - - - - - - - - - - - -

The cork is produced by a lateral meristem or cambium, which would increase the diameter of the tree if not removed.

18. The apical and lateral meristems constitute the chief growth regions of almost all seed plants, but a third type, called the intercalary meristem, may occur in a few plants and function as an important growth region.

As the term intercalary implies, this meristem, in contrast to an apical meristem, is intercalated or inserted between mature tissues.

The stem of a horsetail plant is noticeably jointed in appearance. The stem tissues in the regions between the joints are mature, yet the stem increases in length at the joints. Which of the three meristems, apical, lateral, or intercalary, would you expect to be functioning?

\- - - - - - - - - - - - - - - - -

If new growth occurs between mature tissues, it must be due to isolated intercalary meristems.

19. To visualize the relationship of different kinds of meristems to the structure of the plant, identify the meristems labeled in the drawing on the next page as lateral or apical. (Cover the answers below while you write yours, then refer back to check your answers.)

\- - - - - - - - - - - - - - - - -

a. apical; b. lateral; c. apical

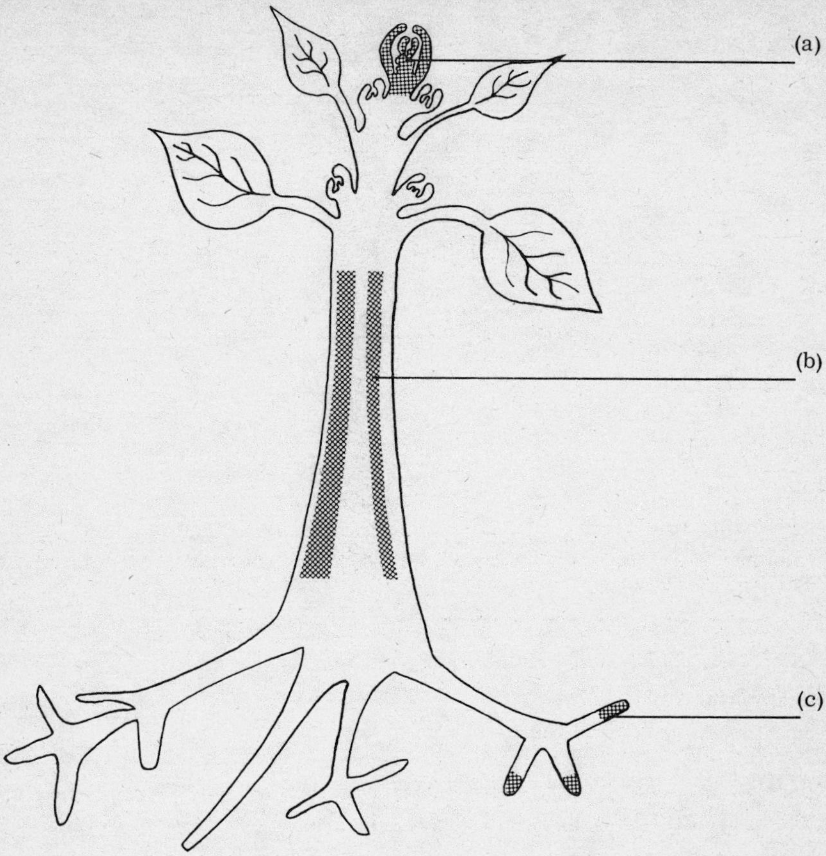

20. The apical meristems produce the first or original tissues in roots, stems, and leaves. Such tissues are spoken of as primary tissues.

If the stem of a house plant becomes too tall and spindly, we can prevent additional primary stem tissue from forming by pinching off the stem tip, thus removing the ________________________________.

\- - - - - - - - - - - - - - - - -

apical meristem

21. Increase in the diameter of the axis results from the deposition of additional tissues formed from the lateral meristem upon the primary tissues. The tissues thus formed from lateral meristems or cambiums are designated as secondary tissues. The simplified diagram on the next page illustrates this generalization.

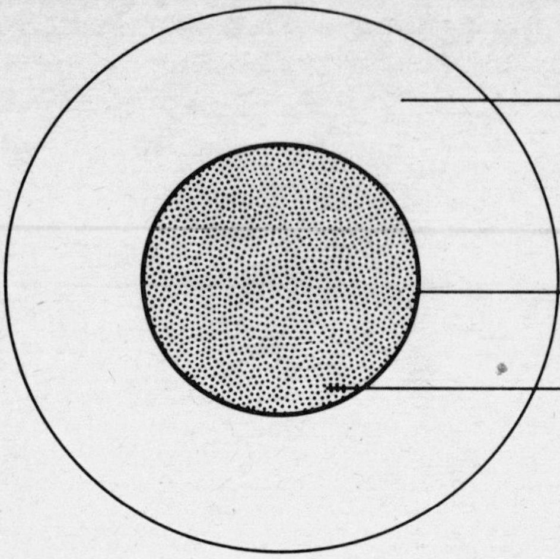

If the cork tissue mentioned in frame 17 is produced by the action of a lateral meristem, it follows that cork is (primary/secondary—choose one) ______________________ tissue.

- - - - - - - - - - - - - - - - -

secondary

22. Identify each of the following meristems by matching the proper term on the right with its description on the left.

________ a. Meristem which typically forms secondary tissue.		1. apical meristem
________ b. Meristem responsible for increase in root length or plant height.		2. intercalary meristem
________ c. Meristem responsible for increased diameter of a tree trunk.		3. lateral meristem
________ d. Meristem inserted between mature tissues.		
________ e. An alternative name is cambium.		
________ f. Meristem forming primary tissue.		
________ g. Meristem forming cork layers in the cork oak.		

- - - - - - - - - - - - - - - - -

a. 3; b. 1; c. 3; d. 2; e. 3; f. 1 (and probably 2); g. 3

23. Now, summarize your understanding of the location and function of apical, lateral, and intercalary meristems by completing the following table.

	location	function
apical	______________________	______________________
lateral	______________________	______________________
intercalary	______________________	______________________

- - - - - - - - - - - - - - - - -

location	function
stem and root tips	stem and root elongation; formation of primary tissues
ring or cylinder of tissue near perimeter of the plant axis	increase in stem and root diameter; formation of secondary tissues
inserted between mature tissues as at nodes of the stem	increase in length of plant axis; formation of primary tissue

24. As plant tissues originate from their meristems and increase in size, the cell walls and general features assume the characteristics of the various mature tissues. These are the permanent tissues which were discussed in frame 14.

Most permanent tissues are simple tissues because they are composed of essentially one kind of cell. The remaining frames of this chapter are devoted to the description of simple permanent tissues.

If you examine microscopically a small piece of a mature Irish potato and find cells of uniform diameter whose function seems to be storage of starch, what conclusions can you make about the composition of that piece of potato? ______________________________________

____

- - - - - - - - - - - - - - - - -

You should conclude that it is a tissue because it is composed of a group of cells which are functioning as storage units, and you should also conclude that the tissue is simple, since there is only one type of cell present.

25. Let us consider the chief simple permanent tissues which make up a typical plant. Parenchyma (pa-renk´-e-ma) is the most common and perhaps the least specialized or differentiated of all the tissues and occurs throughout the plant.

Much of the food and water stored by a plant is found in parenchyma, so that most of the plant parts we eat are composed of that tissue. When you eat an apple which has been peeled and cored, you are eating mostly parenchyma.

Keeping in mind that watermelons contain a great deal of water and some sugar, why should you expect the edible portion of the watermelon to be composed chiefly of parenchyma? ____________________

____

- - - - - - - - - - - - - - - -

Parenchyma is a storage tissue for food and water.

26. In addition to serving as a storage tissue, parenchyma performs other functions. Although classed as a permanent tissue, parenchyma may revert to a meristematic condition under certain circumstances and produce new cells to repair wounds in the plant. Several other plant activities may occur because of meristematic parenchyma cells.

In addition to its other roles, parenchyma tissue often develops the green pigment, chlorophyll, and functions as a food-producing tissue.

To summarize the importance of parenchyma, list three general functions which parenchyma may perform in the plant. ____________

____

____

- - - - - - - - - - - - - - - -

storage of food and water; meristematic activity; food manufacture (in any order)

27. Parenchyma cells may present a varied appearance when viewed under the microscope, but a typical cell has dimensions which are essentially as wide as long (isodiametric) and an active protoplast enclosed by a thin primary cell wall, the chief constituent of which is cellulose.

In the photograph shown on the next page, which of the cells is/are isodiametric? ____________________

- - - - - - - - - - - - - - - -

All of the cells appear to be isodiametric.

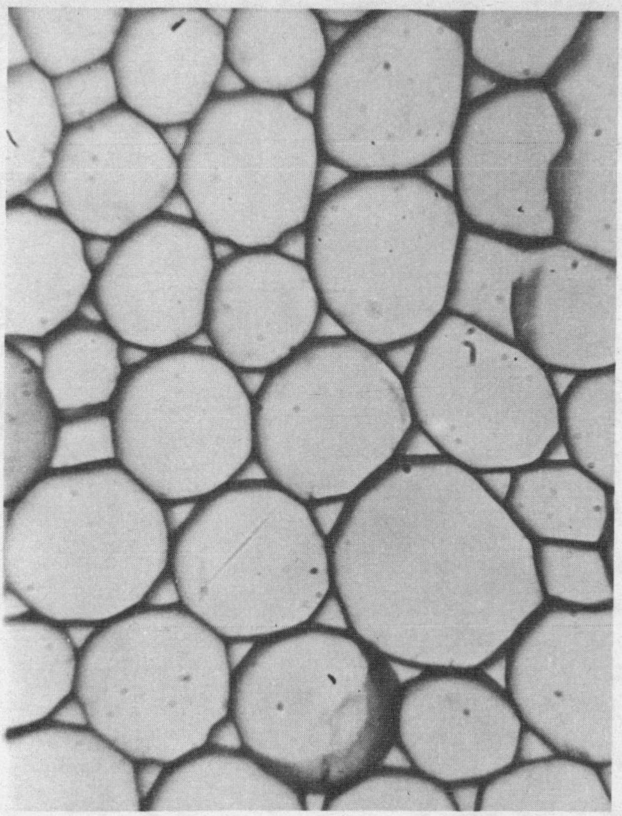

28. The individual cell walls of parenchyma may separate from each other in certain areas and form intercellular spaces as the tissue matures (see photograph on next page).

Although intercellular spaces are typical of parenchyma, they occur in other tissues as well. A loosely organized arrangement with spaces between the cells allows for a free circulation of oxygen and other gases throughout the plant.

Keeping in mind that cells may be artificially separated from each other by maceration in strong acids, what cementing structure separating adjacent cell walls would you expect to be directly involved in the formation of intercellular spaces? ______________________________

____

- - - - - - - - - - - - - - - - -

The middle lamella has been dissolved to allow the primary walls of the cells to separate from each other in specific areas.

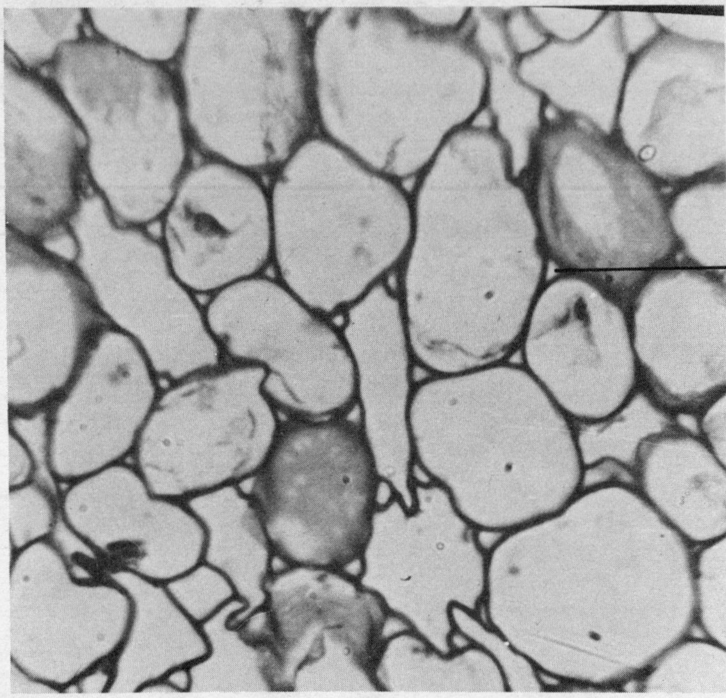

29. A young plant whose permanent tissues are just emerging from the apical meristems may not be capable of growing erect without mechanical support. This support is often provided by collenchyma (col-lenk´-e-ma) tissue. Generally located in the peripheral portion of the stem and in various parts of the leaf, the flexible and plastic cell walls of collenchyma provide adequate support for its neighboring cells. Since collenchyma seldom produces secondary walls, it represents cells with extensively thickened primary walls. Collenchyma is living tissue and thus is capable of manufacturing food or becoming meristematic to produce wound tissue.

Nonwoody plants growing beside a busy highway are subjected to much wind produced by passing cars. Anatomical examination of the stems of the plants generally reveals a large amount of collenchyma present. Of what advantage is the presence of this tissue to these plants?

____

____

- - - - - - - - - - - - - - - - - -

The stems possess great flexibility and can withstand the whipping action of the wind without sustaining injury.

30. A microscopic view of a cross section of collenchyma cells reveals primary cell walls which are generally unequal in thickness, as illustrated on the next page.

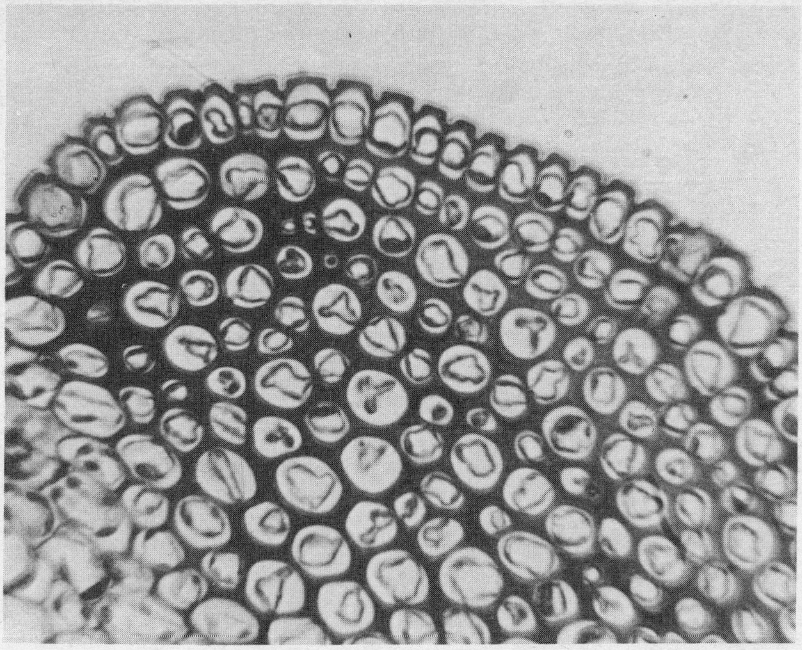

cross section of collenchyma

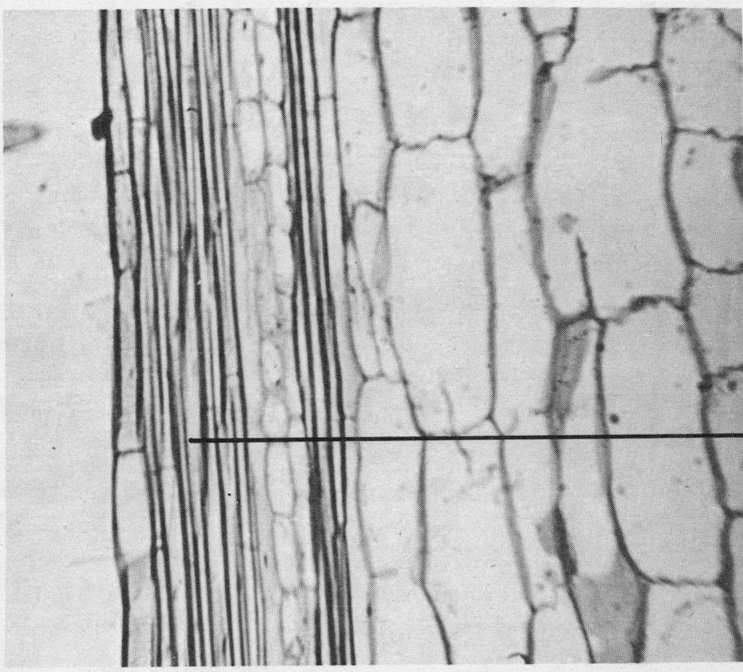

longitudinal section of collenchyma

The elasticity of these walls gives meaning to the literal translation ("glue tissue") of the word collenchyma.

Collenchyma physically resembles parenchyma in that both possess cells with living and active protoplasts surrounded by primary walls. They differ in the thickness of the walls and in the greater length of collenchyma cells in contrast to the isodiametric cells of parenchyma.

The close relationship between the two tissues is apparent in stems where collenchyma and parenchyma lie adjacent to each other. In many instances there is no definite line of demarcation between them, because cells having walls with intermediate thickness occur between the two distinct kinds of tissue.

Indicate the similarities and differences between parenchyma and collenchyma by matching the proper tissues with their descriptions below.

________ a. Unequally thickened walls.

________ b. Living tissue.

________ c. Primary cell walls.

________ d. Isodiametric cells.

________ e. Capable of meristematic activity.

________ f. Occurs in plants in larger amounts.

1. parenchyma
2. collenchyma
3. both collenchyma and parenchyma

- - - - - - - - - - - - - - - - - -

a. 2; b. 3; c. 3; d. 1; e. 3; f. 1

31. As the plant matures, other tissues besides collenchyma provide mechanical support for the young plant. One such type of supportive tissue is the simple and permanent <u>sclerenchyma</u> (skli-ren´-ki-ma), whose name literally means "hard tissue." Occurring in all parts of a plant, sclerenchyma may form the familiar stringy tissues in some of our plant foods such as celery stalks or the hard coverings of nuts and peach pits.

The individual cells of sclerenchyma vary in appearance, but they are characterized by thick secondary walls whose chemical nature and physical thickness impart a greater degree of hardness to sclerenchyma tissue than that of collenchyma. Neither parenchyma nor collenchyma cells typically have secondary walls. At the same time, however, sclerenchyma retains its flexible nature. Although the secondary wall is chiefly cellulose, hardening of sclerenchyma tissue is largely due to the deposition of another substance called <u>lignin</u> which impregnates the middle lamella, primary, and secondary cell walls. Many times sclerenchyma cell walls become so greatly thickened that little room is left for the protoplast. Consequently, the protoplast disappears from these cells upon maturity and they function as dead cells.

Do you think that sclerenchyma would be a suitable substitute tissue for collenchyma in a very young plant? Give the reason for your answer.

____

\- - - - - - - - - - - - - - - - - -

Sclerenchyma would not be an efficient tissue in a young plant whose cells are still elongating because sclerenchyma is harder and therefore less pliable than collenchyma.

32. The cells which comprise the supportive tissue sclerenchyma may assume two forms: (1) *fibers*, and (2) *sclereids*. They differ from each other chiefly in form or shape. A typical fiber is greatly elongated with sharply pointed end walls, and as a form of sclerenchyma tissue, it possesses a greatly thickened secondary wall. Mature fibers typically lack protoplasts but when they do exist, the protoplasm is relatively inactive. The following photograph depicts the nature of a fiber.

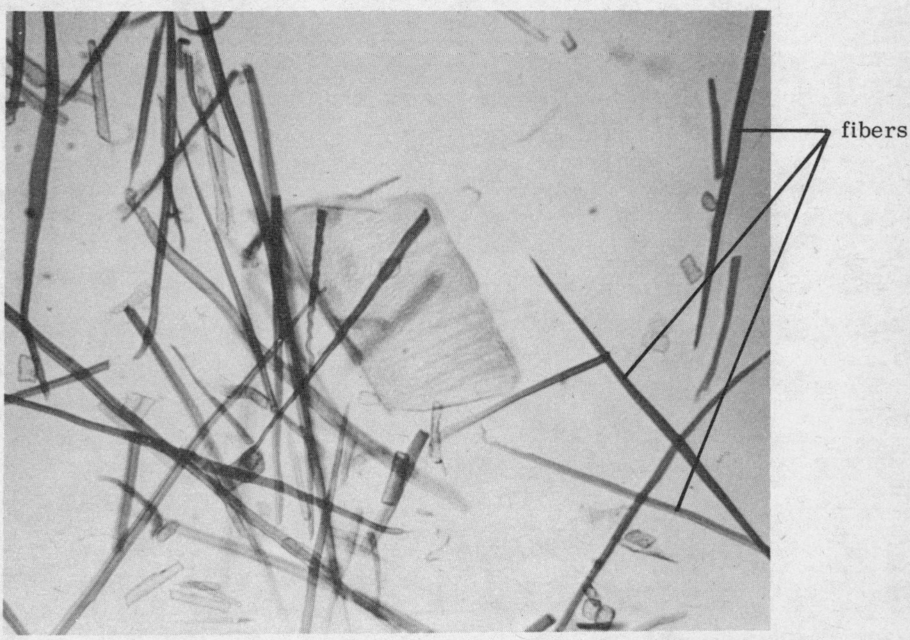

Fibers may occur singly or in groups in all parts of a plant. Vertical series of fibers with their end walls overlapping produce fibrous strands in the plant and they are frequently used commercially in fashioning rope and various types of cordage.

Check the following statement(s) which accurately describe(s) the nature of fibers.

________ a. Fibers may function as dead cells.

________ b. Fibers often closely resemble parenchyma.

________ c. The thick walls of fibers are produced while the fiber is still growing.

________ d. Fibers may be found in all parts of the plant.

- - - - - - - - - - - - - - - - - -

a and d are true statements

33. Some parts of a plant may be supported or protected by sclereids. Microscopic sclereids are shorter in length than fibers which are often visible to the naked eye. Sclereids may even appear isodiametric in many instances, and in contrast to fibers, they may branch to form strange and interesting shapes (see photographs).

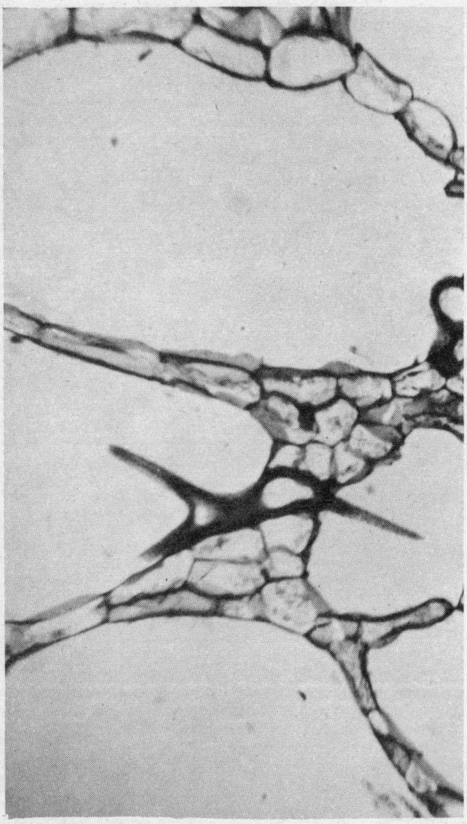

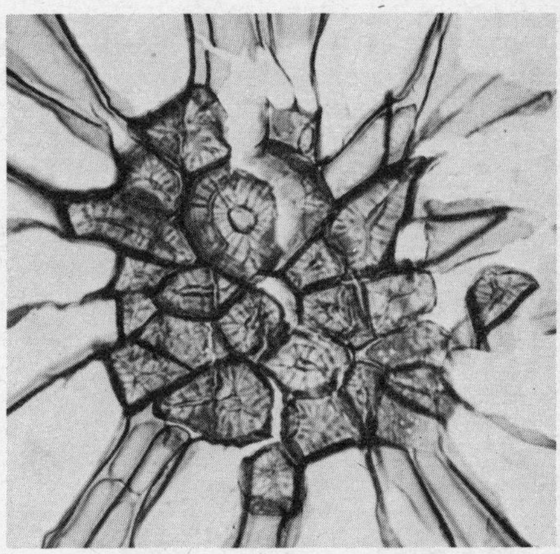

isodiametric sclereids from pear fruit (Courtesy Carolina Biological Supply Company.)

branching sclereid from an oleander leaf

a. Fibers and sclereids are two forms of what kind of tissue?

____

b. How do fibers and sclereids differ in shape? ________________

____

- - - - - - - - - - - - - - - - -

a. sclerenchyma
b. Fibers are slenderly elongated with sharply pointed end walls; sclereids are shorter in length and often branch freely.

34. The primary wall of a young sclereid, like that of almost all other cells, possesses a number of pits. Scarcely noticeable when primary walls remain thin, pits become prominent features in a greatly thickened secondary wall because the additional wall material also fails to be deposited in the pit areas. The pits in many sclereids branch to form interesting patterns in their walls as illustrated below.

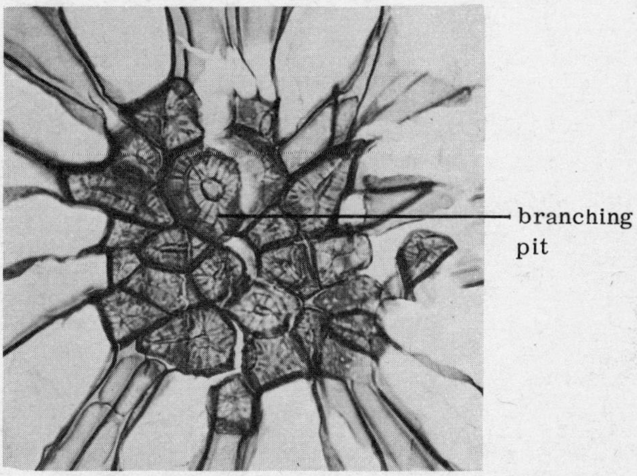

sclereids with branching pits (Courtesy Carolina Biological Supply Company.)

Why would you expect the pits in sclereids to be far more prominent than those in the walls of most parenchyma cells? ________________

____

____

- - - - - - - - - - - - - - - - -

The primary cell walls around parenchyma cells are generally thin, thus the pits are less noticeable than in the thicker secondary walls which are characteristic of sclereids.

35. The extremely hard layers of tissue in some fruits and seeds are composed of various forms of sclereids, but the isodiametric sclereids with

prominent branching pits are most often encountered. These sclereids are called stone cells.

When examined under the microscope, the flesh of a pear shows many isolated clusters of sclereids with equal diameters and prominent branching pits. The gritty nature of pear flesh, then, is due to the presence of a type of sclereid called ______________________.

- - - - - - - - - - - - - - - - - -

stone cells

36. Collenchyma, sclereids, and fibers support the plant body. Summarize the similarities and differences between these tissues by matching the cells with their characteristics listed below. Some blanks may require more than one answer.

________ a. Cells often branched.		1. fibers
________ b. Secondary walls present.		2. collenchyma
________ c. Cell walls plastic to allow for stretching.		3. sclereids
________ d. May function as dead cells.		
________ e. Often possess commercial value.		

- - - - - - - - - - - - - - - - - -

a. 3; b. 1 and 3; c. 2; d. 1 and 3; e. 1

37. Now let us consider epidermis as another type of simple living permanent tissue. The epidermis provides a protective covering over all external parts of a plant derived from the apical meristems (the primary plant body).

As is true of almost all tissues, the individual cells of the epidermis may vary in appearance. Some cells are relatively unspecialized and resemble pavement blocks (see photograph on next page), and others are more specialized in appearance and function.

With few exceptions, the epidermal tissue forms a layer one cell thick over the plant body. The cell walls become impregnated with cutin, a fatlike material secreted by the protoplasts, which renders the walls somewhat impervious to water. Cutinization is usually heaviest on the outside walls, and an extensive production of cutin builds a waxy cuticle over the entire epidermal surface, as the second photograph on the next page shows.

Cutinization of the epidermis also retards excessive water evaporation from the moist underlying tissues.

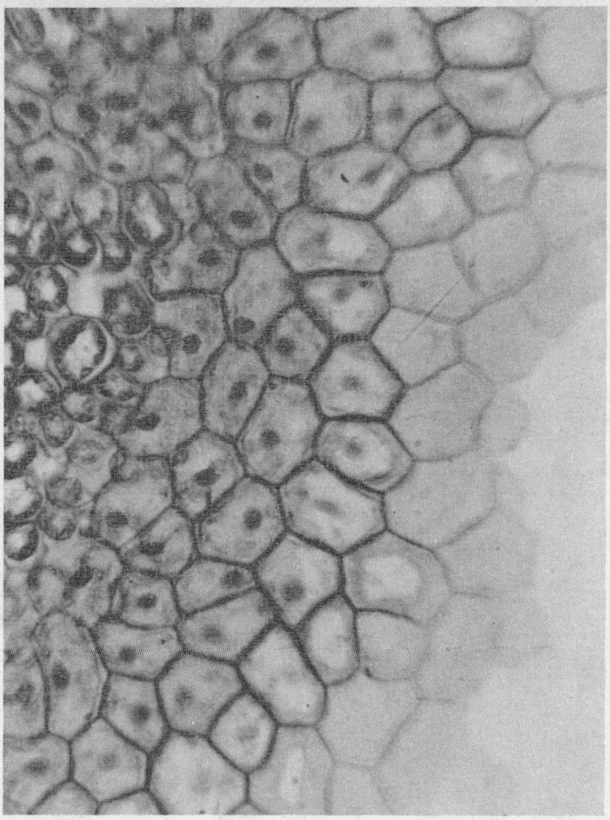

leaf epidermis (surface view)

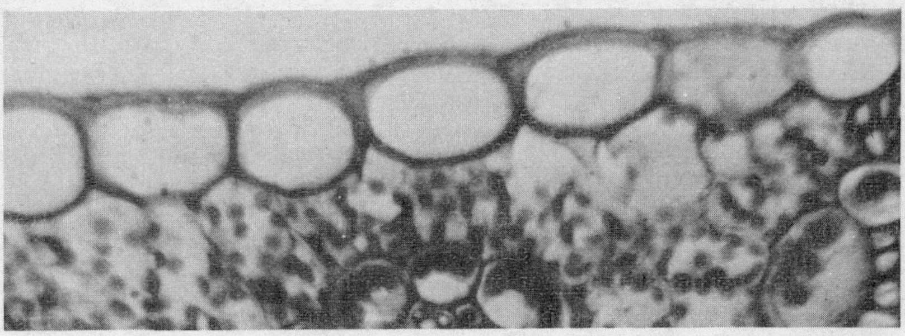

leaf epidermis (cross section)

What structure would you guess is involved when you "polish the apple"? ______________________________

- - - - - - - - - - - - - - - - - -

The cuticle over the surface of the apple takes on a high luster when rubbed.

38. Retardation of the loss of water vapor through the epidermis is a desirable function of that tissue, but an epidermis completely impervious to water vapor would not allow the necessary exchange of oxygen and carbon dioxide to take place between the living underlying cells of the plant and the atmosphere.

The two photographs in frame 37 illustrate the nature of epidermal cells. The surface view of the leaf epidermis depicts flattened cells lying closely together without intercellular spaces separating them. While this is an accurate representation of the appearance of most epidermal cells, the occurrence of guard cells in the epidermis of almost all aerial (above ground) plant parts makes gaseous exchange possible across the epidermis.

The guard cells in the epidermis allow the exchange of the vital gases, ______________________ and ______________________, between the underlying tissues and the atmosphere.

- - - - - - - - - - - - - - - - -

oxygen; carbon dioxide (in either order)

39. Guard cells are special crescent-shaped cells which occur in pairs in scattered arrangement throughout the epidermis. They are oriented so that their concave surfaces face each other. The large intercellular space formed between the two cells bears the name stoma (plural, stomata), from the Greek word for mouth. Some authors find it more convenient to use the term to refer to the entire apparatus of guard cells and space (see photograph on next page).

Guard cells lose their crescent shape when water leaves the cells, as frequently happens. The variation in shape of the guard cells affects the size of the stomatal openings which, in turn, regulates the amount of water vapor, oxygen, and carbon dioxide which passes through the stomatal opening.

A preparation on the market gives a bright luster to leaves which are coated with it. Many botanists do not approve of its use. Can you guess why? ____

____

- - - - - - - - - - - - - - - - -

The stomata on the leaf surfaces may become clogged, thus interfering with the vital exchange of gases across the epidermis.

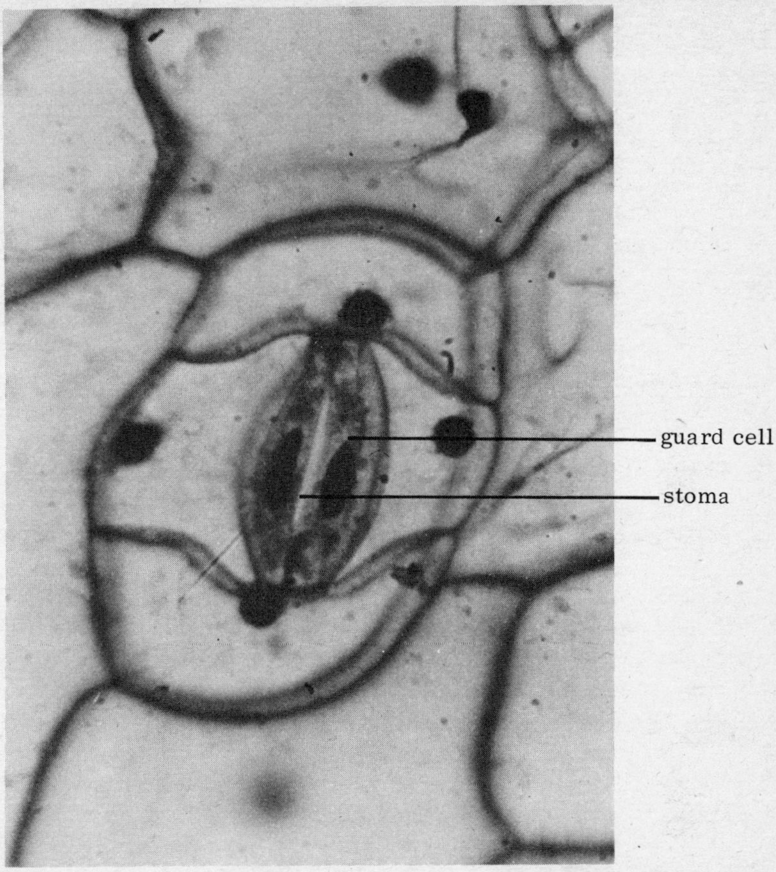

40. Many epidermal cells form various extensions which appear as hairs or scales on the surface of the plant. The long lateral extensions from epidermal cells of the root are responsible for uptake of water by that organ, as we shall discuss later. The hairy appearance of plant stems and leaves is due to epidermal appendages.

As Eli Whitney knew, cotton fibers grow from the surface of seeds, and in processing cotton the fibers and seeds must be separated. From the information presented in this frame, can you suggest the anatomical origin of cotton fibers? ______________________________

____

- - - - - - - - - - - - - - - - -

Cotton fibers are outgrowths from epidermal cells of the cotton seed. Cotton fibers should not be confused with the fiber cells of sclerenchyma.

41. One tissue remains to be discussed in our consideration of simple permanent tissues. Cork, unlike the previous tissues discussed, arises from the activity of a lateral meristem or cambium. Therefore, cork

is secondary tissue in origin, but it is classified as simple and permanent.

The function of cork is a simple one; it is protective in nature and replaces the epidermis in those plants which produce extensive secondary tissues.

A unique feature of a corn plant is that all of its tissues arise from apical meristems yet the diameter of its stem increases significantly during the summer. Would you expect to find cork or epidermis at the surface of the stem? Justify your answer. ______________________

____

____

- - - - - - - - - - - - - - - - -

You would expect to find an epidermis because extensive cork is formed from a lateral meristem. Increase in the diameter of the corn stem results from the enlargement of individual existing cells.

42. Since you are familiar with cork tissue because of its many commercial uses, you may be able to guess how the tissue will appear when viewed under the microscope.

The waxy feel of bottle cork reflects the nature of its thick secondary cell walls which become impregnated with <u>suberin</u>, a fatlike substance chemically related to cutin. The characteristic brown color of cork results from various brown pigments absorbed by its walls. Cork cells die upon maturity and the cell cavities are usually filled with air, thus explaining the buoyancy of cork tissue. The dead cork cells, highly impermeable to water, appear arranged in radial rows under the microscope because of the manner in which they arise from the lateral meristem. The photograph on the next page represents the appearance of typical cork tissue.

The term <u>cell</u> was first used by Robert Hooke in 1665 when he examined a thin section of bottle cork under his primitive microscope. What part of the cork cells was Hooke able to see? ____________________

____

- - - - - - - - - - - - - - - - -

He saw only cell walls. The protoplasts were missing because cork cells are dead upon maturity.

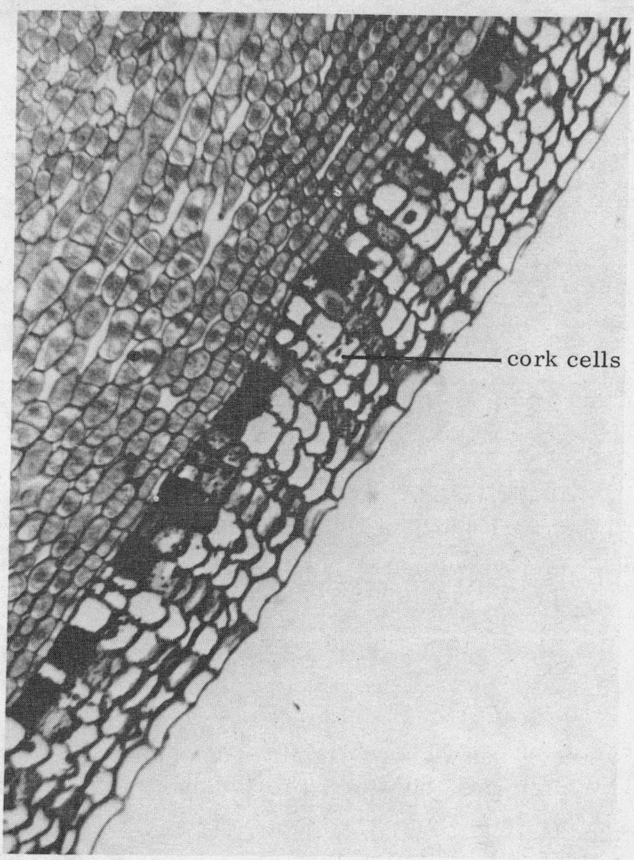

43. To summarize our discussion of simple plant tissues, match the name of the tissue(s) with its description. Some blanks may require more than one answer.

________	a. Characteristic nonliving fiber cells providing the plant with mechanical support.	1. parenchyma
________	b. Tissue providing protection from water loss.	2. collenchyma
________	c. Tissue having noticeable pits in its cell walls.	3. sclerenchyma
________	d. Only tissue derived entirely from lateral meristem.	4. epidermis
________	e. Tissue capable of functioning meristematically and in food manufacture.	5. cork
________	f. Peach "fuzz" is probably a part of what tissue?	

(continued)

________ g. Tissue characteristically only one cell layer thick.

________ h. Cell walls contain waxy suberin.

________ i. Cells may function in food manufacturing.

________ j. Tissue made up in part of guard cells.

________ k. Functions primarily in storage of food and water.

- - - - - - - - - - - - - - - - - -

a. 3; b. 4 and 5; c. 3; d. 5; e. 1 and 2; f. 4; g. 4; h. 5; i. 1 and 2; j. 4; k. 1

44. Tissues such as parenchyma, made up almost entirely of one kind of cell, are called (primary/simple) ____________________ tissues.

- - - - - - - - - - - - - - - - - -

simple

45. One simple tissue that we have examined can be made up of either of two cell types. Both forms, however, provide mechanical strength and protection to the plant. The name of this tissue is ____________________. Name the two kinds of cells in this tissue, and describe them. ________

____

____

____

- - - - - - - - - - - - - - - - - -

sclerenchyma
fibers and sclereids. Fibers are elongated cells with pointed end walls while sclereids appear irregularly branched or occasionally isodiametric.

46. Cork, which is formed by a lateral meristem or ____________________, is an example of a (primary/secondary) ____________________ tissue, although it is still regarded as a simple tissue.

- - - - - - - - - - - - - - - - - -

cambium; secondary, because all tissues derived from a lateral meristem are by definition secondary tissues

47. Why is cork regarded as a simple tissue? ______________________________

____

Why is it considered to be a secondary tissue? ____________________

____

- - - - - - - - - - - - - - - - -

It consists almost entirely of only one kind of cell.
It is formed by a lateral rather than an apical meristem.

48. This chapter has discussed the nature of the following mature tissues: parenchyma, collenchyma, sclerenchyma, epidermis, and cork. What single characteristic is shared by these tissues? ____________________

____

- - - - - - - - - - - - - - - - -

They are all simple tissues since they are composed of only one type of cell.

SELF-TEST

Before you proceed, take this short Self-Test to help you reinforce what you have learned in this chapter. All questions are to be answered in the light of information in Chapter 1 of this book. Answers are given at the end of this quiz.

1. Diagrammed below are several plant cells. Write in the items identified by letter by selecting the number of the proper terms from among those listed below the figure.

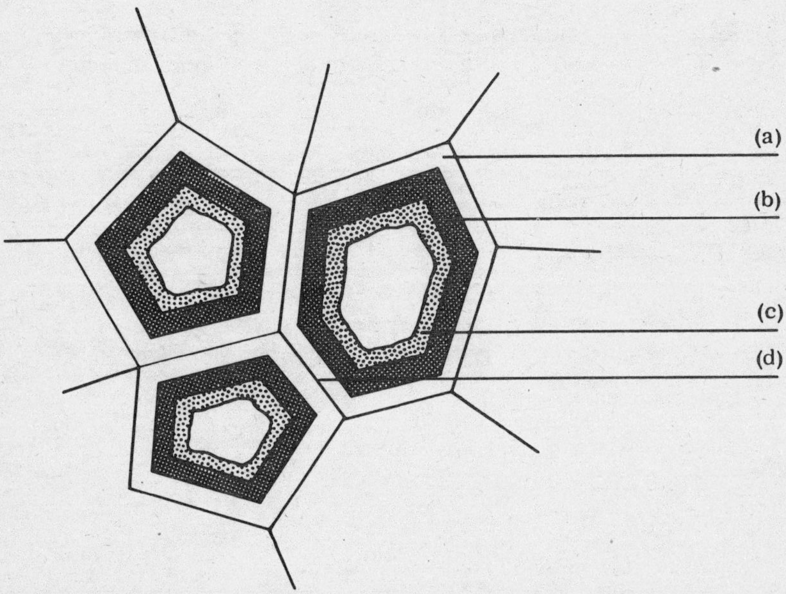

1. cambium	4. apical meristem
2. secondary wall	5. protoplast
3. middle lamella	6. primary wall

2. Which of the four parts illustrated in the diagram in question 1 are living? ______________________________

3. The presence of starch grains in a large group of cells, as in a potato, indicates that they are performing the specific function of food storage. We may call this group of cells which have a common function a(n) ______________________________.

4. Place the following in order of decreasing complexity, beginning with a part that is visible to the unaided eye.

   (1) cells, (2) organs, (3) protoplasm, (4) tissues

   ____

5. Secondary tissues of the plant are formed by what kind of meristem? (Give both names for this meristem.) ______________________

   Primary tissues of the plant are formed by what kind of meristem?

   ____

6. An increase in the height of a plant results from the activity of a(n) ______________________ meristem, while increase in plant diameter results from activity of a(n) ______________________ meristem.

7. To help you clarify the similarities and differences between three important simple plant tissues, complete the following matching exercise.

   ________ a. Cells have thick, hard secondary walls.

   ________ b. Living tissue with active protoplasts providing mechanical support.

   ________ c. Cells contain chlorophyll and can manufacture food.

   ________ d. Cells generally lack protoplasts at maturity.

   ________ e. Living isodiametric food-storing cells.

   ________ f. Makes up the hard coverings of nuts.

   ________ g. Capable of meristematic activity.

   ________ h. Provide mechanical support.

   ________ i. Principal function is storage of water and storage and manufacture of food.

   1. parenchyma
   2. collenchyma
   3. sclerenchyma
   4. both parenchyma and collenchyma
   5. both collenchyma and sclerenchyma

8. We hear a great deal about the potentially harmful effects of the misuse of marijuana. However, one of the beneficial uses of the plant from which marijuana is obtained is the rope or cord that can be made from the fibers in its stem. We call this material "hemp." We might reasonable expect hemp to be a form of what simple plant tissue? ______________________

9. Cork and epidermis are two tissues that may be found on the surface of a plant. Compare and contrast these two tissues by completing the matching exercise on the next page.

________ a. A primary simple tissue.

________ b. Formed by a lateral meristem.

________ c. Contains a waxy substance in its cell walls.

________ d. Includes special structures called guard cells and stomata.

________ e. Cells are dead at maturity.

1. cork
2. epidermis
3. both cork and epidermis

10. Summarize your knowledge of simple plant tissues by matching the names of the different kinds of tissues with the descriptive statements. Some blanks may require more than one answer.

________ a. Typically made up of living cells.

________ b. Generally capable of becoming meristematic.

________ c. Provide mechanical support to plant parts.

________ d. Because of location and cell wall composition, these tissues are impervious to water and prevent water loss by the plant.

1. parenchyma
2. collenchyma
3. sclerenchyma
4. epidermis
5. cork

Answers

Compare your answers to the Self-Test with the correct answers given below. If all of your answers are correct, you are ready to go on to the next chapter. If you missed any questions, review the frames indicated in parentheses following the answer before you go on.

1. a. 6 (frame 5)
   b. 2 (frame 8)
   c. 5 (frame 4)
   d. 3 (frame 6)

2. protoplast (frames 1 and 4)

3. tissue (frame 10)

4. 2—4—1—3 (frames 1-13)

5. lateral meristem or cambium; apical meristem (frames 14-23)

6. apical, lateral (frames 14-23)

7. a. 3 (frame 31)
   b. 2 (frames 29-30)
   c. 4 (frames 26 and 29)
   d. 3 (frame 31)

e. 1 (frames 25-27)
f. 3 (frame 31)
g. 4 (frames 26 and 29)
h. 5 (frames 29 and 31)
i. 1 (frame 25)

8. sclerenchyma (frame 32)

9. a. 2 (frame 37)
b. 1 (frame 41)
c. 3 (frames 37 and 42)
d. 2 (frames 38 and 39)
e. 1 (frame 42)

10. a. 1, 2, and 4 (frames 26, 29, 37)
b. 1 and 2 (frames 26 and 29)
c. 2 and 3 (frames 29 and 31)
d. 4 and 5 (frames 37 and 42)

CHAPTER TWO

Complex Plant Tissues: Xylem and Phloem

1. Not all plant tissues are composed of the uniformly basic cell types which we discussed in Chapter 1. Microscopic examination of certain tissues reveals that they are composed of several different kinds of cells. These tissues may be referred to as complex tissues, to distinguish them from simple tissues.

   In what way are simple and complex tissues alike, and in what way do they differ? ____

   ____

   ____

- - - - - - - - - - - - - - - - -

Both simple and complex tissues are composed of cells which work together in performing a specific function. Simple tissues are composed of only one kind of cell, while complex tissues are composed of several different kinds of cells.

2. Two complex tissues, xylem and phloem, function in the plant as conductive tissues. They are responsible for the movement of water and soluble foods through the plant. Although separate and distinct tissues, they are more often referred to collectively as vascular tissue to reflect their conductive roles. Because of their complexity and importance, this chapter is concerned solely with these tissues.

   Two individual kinds of tissues comprise the vascular tissue. They are ____________________ and ____________________. They are described as complex tissues because ____________________

   ____.

- - - - - - - - - - - - - - - - -

xylem, phloem (either order); they consist of several different cell types

3. Let us consider the composition and general function of xylem tissue. Most of the water required by a plant comes from the soil. Soil water is absorbed by the roots, and the unique features of the xylem tissue enable water to be translocated to all aerial parts of the plant. Therefore, the general direction of water movement is upward from the roots to the upper portions of the plant.

On a bright warm summer day, a tree may lose a surprisingly large amount of water from its leaves. The leaves of the tree wilt unless the water content is replenished immediately.

Can you guess the source of the replenished water and how it can reach to the top branches of the tree? ______________________________

____

- - - - - - - - - - - - - - - - -

The water comes from the soil and the xylem tissue translocates the water.

4. You may conclude that xylem possesses tubular types of cells to allow such rapid movement of water through it, and a microscopic examination of the tissue proves that this is, indeed, the situation.

An important water-conducting unit in xylem is the tracheid, a narrow elongated cell with tapered and somewhat pointed end walls. The thick secondary walls usually become impregnated with lignin (see Chapter 1, frame 31). The protoplast dies upon maturity so that a tracheid functions as a dead hollow cell with a large cavity or lumen (see photograph).

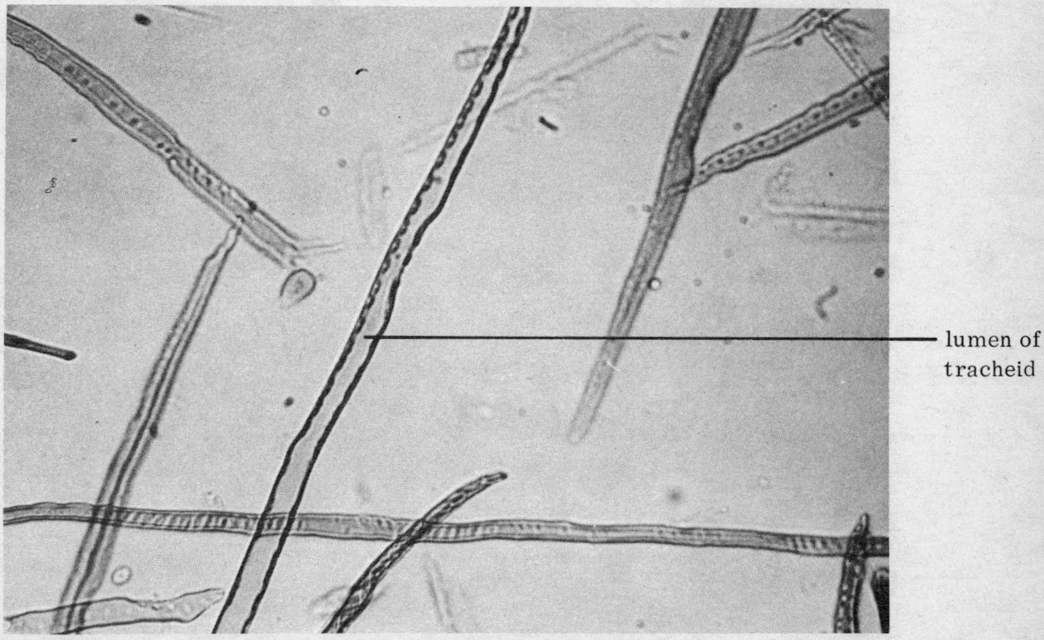

longitudinal view of a tracheid

Tracheids are arranged in a vertical series so that the tapered end wall of one cell overlaps that of the next tracheid above it. Large pits in the overlapping walls allow for movement of water upward through the tracheids.

The thick lignified walls of tracheids, with the overlapping arrangement, suggest a secondary and less important role of tracheids besides water conduction. Can you guess what that role is? ________________

____

- - - - - - - - - - - - - - - - - -

Like fibers, which they resemble, tracheids furnish some mechanical support to the plant body.

5. Tracheids function so efficiently as water-conducting cells that many vascular plants such as the pine tree have xylem composed solely of tracheids. Flowering plants, however, possess vessel members as well as tracheids for water conduction.

Vessel members, like tracheids, are dead cells with thick lignified secondary walls, but the end walls of vessel members disappear as the cells mature. Therefore, a vertical series of vessel members are not separated from each other by cross walls, as illustrated below.

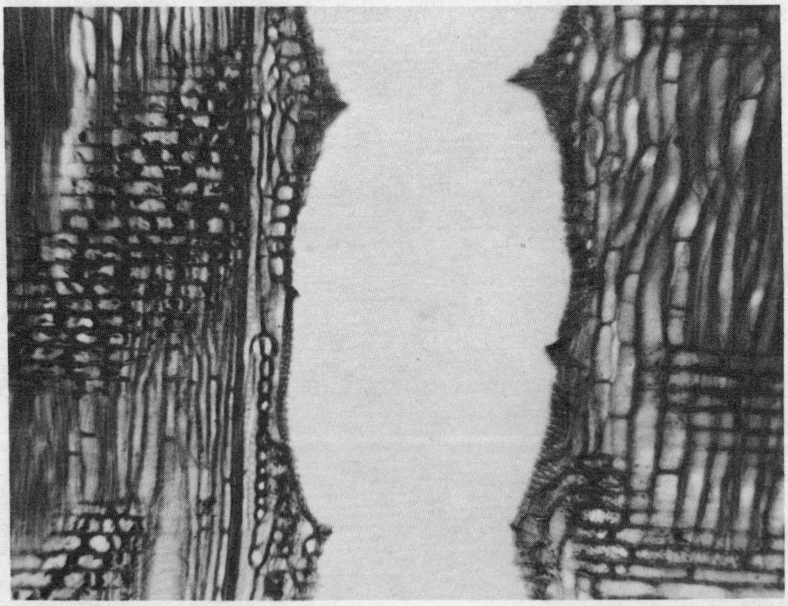

longitudinal view of a vessel

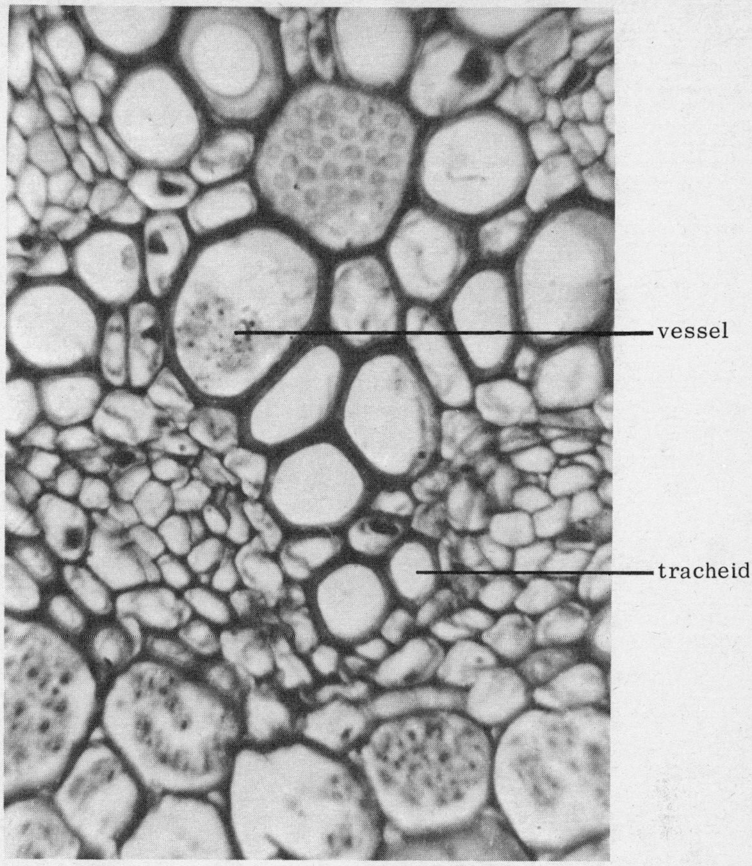

cross section of xylem tissue

When tracheids and vessels are viewed in cross section under the microscope, they appear quite similar, as the cross section photograph shows. How can you distinguish between them if you are given both cross and longitudinal xylem sections to examine? ______________________

____

- - - - - - - - - - - - - - - - - -

The longitudinal section should reveal the absence or presence of cross walls. The pointed end walls would, of course, be those of tracheids.

6. As the longitudinal view photograph in frame 5 depicts, a vertical series of vessel members forms a tube whose walls are those of the side walls of the dead vessel members. Such tubes composed of nonliving cells are called <u>vessels</u>. The term <u>vessel members</u> is an appropriate term for those cells comprising the vessel.

The xylem tissue of flowering plants may be considered more complex than that of the pine tree. In a flowering plant water is conducted in xylem through two kinds of cells called ________________ and ________________. However, water in the xylem of a pine tree moves solely through ________________.

- - - - - - - - - - - - - - - - -

vessels, tracheids (either order); tracheids

7. We learned in Chapter 1, frame 9 that the secondary wall may be deposited unequally on the primary cell wall. This is especially true of the walls of both tracheids and vessel members.

Secondary walls may be laid down in a cell in the form of rings (annular thickenings) or in a spiral manner (spiral thickenings). Vessel members and tracheids with these patterns are capable of some degree of stretching. Thus, they are often formed in areas where growth and elongation of the surrounding tissues are still occurring (see illustrations below and on next page).

a vessel with annular thickenings

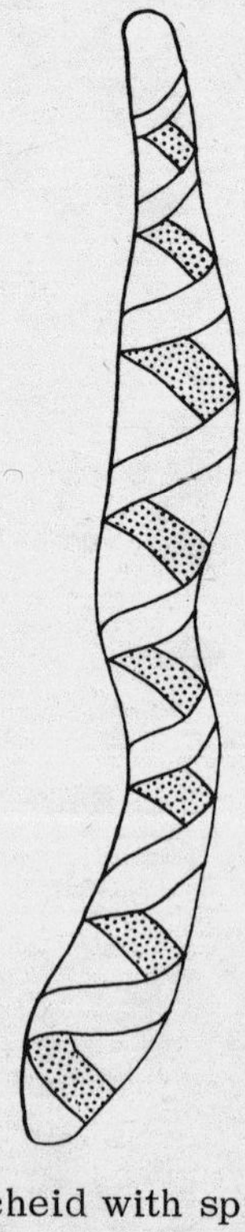

a tracheid with spiral thickenings

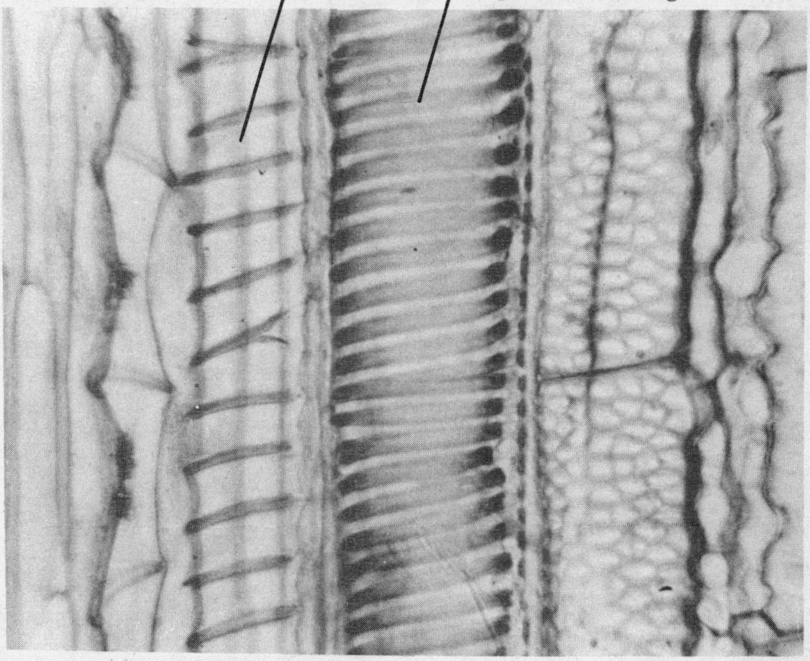

vessels with annular and
spiral thickenings

In tracheids and vessel members of xylem two kinds of wall thickenings are frequently noted, the ringlike __________________ thickenings and ____________________ thickenings. What causes these cell wall thickenings? ____

- - - - - - - - - - - - - - - - - -

annular, spiral; unequal deposition of secondary cell wall material on the inside of the primary wall

8. Some cells have an even greater unequal deposition of secondary wall materials which forms scalariform (ladderlike) thickenings composed of tight spirals with linking bars. Even greater deposition in other cells results in reticulate (netlike) thickenings. When the secondary walls cover most of the primary walls except in certain isolated spots, pits are formed. Look at the diagram on the next page of these three forms of vessel walls.

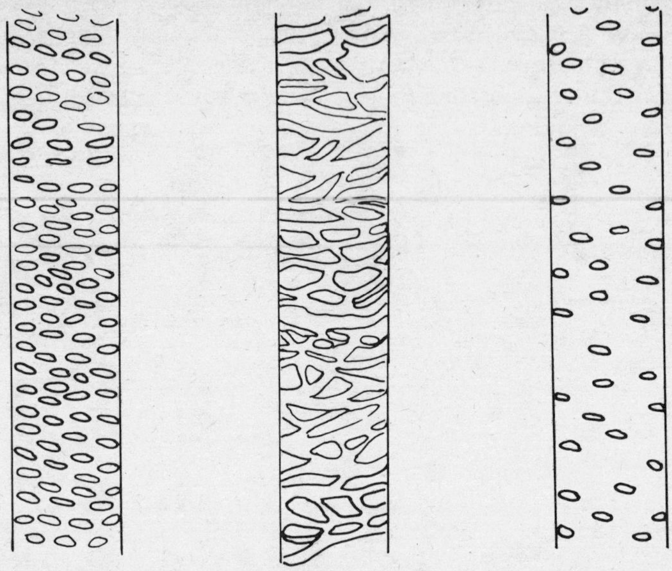

reticulate vessel scalariform vessel pitted vessel

As you may expect, wall deposition often occurs in such a fashion that the types of thickenings blend into one another, thus making accurate classification of wall types impossible.

Check the true statement(s) which describe(s) wall patterns in xylem units.

________ a. Scalariform thickenings are due to unequal secondary wall formations.

________ b. Pits are holes in the primary and secondary cell walls.

________ c. Extensive secondary wall formation renders the xylem unit less capable of stretching.

________ d. Secondary thickenings occur on the walls of both tracheids and vessel members.

- - - - - - - - - - - - - - - - -

a, c, and d are true statements

9. To summarize, the walls of vessels and tracheids are characterized by the presence of rings, spirals, or pits. What is responsible for these distinctive cell wall patterns? ________________________________

____

- - - - - - - - - - - - - - - - -

The unequal deposition of secondary wall material on the primary wall causes these patterns.

10. Many of the pits found in the walls of tracheids and vessel members are simple pits which are formed by the failure of the secondary wall to form over thin areas of the primary wall as described in Chapter 1, frame 7. In addition to simple pits, bordered pits may be formed on the walls of vessel members and tracheids when the secondary wall forms a circular raised overhanging projection over the previously pitted primary wall. Bordered pits present a bizarre and interesting appearance when seen in both surface and side views (see illustrations below and on next page).

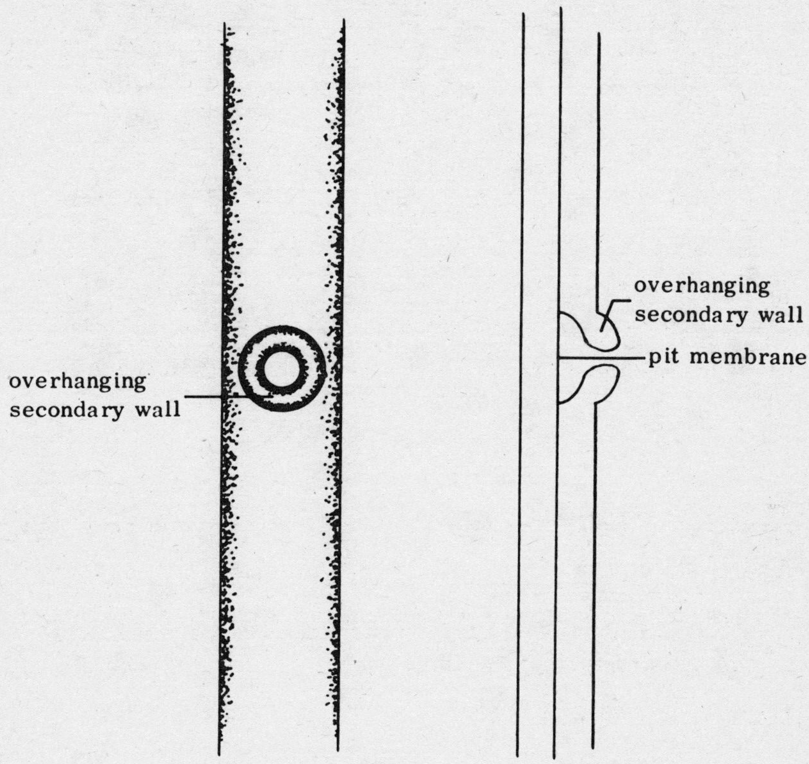

diagrams of bordered pit, surface view (left) and side view (right)

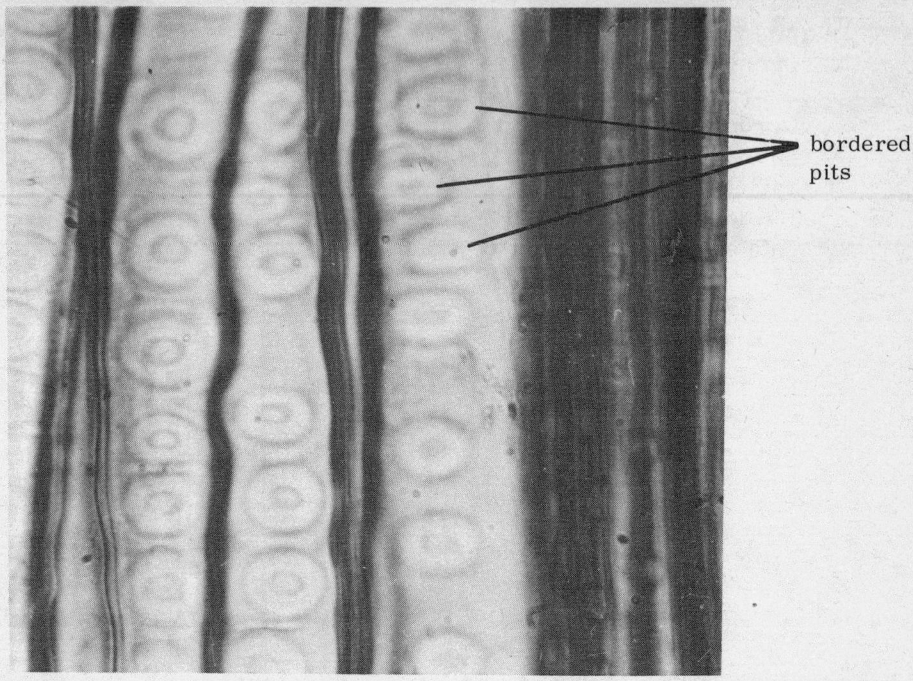

photograph of bordered
pits, surface view

Two kinds of pits may be seen on the walls of tracheids and vessel members. They are ________________ pits and ________________ pits.

- - - - - - - - - - - - - - - - -

simple pits, bordered pits (either order)

11. Although a bordered pit is sometimes formed on only one wall of two adjoining cells, bordered pits usually involve pit pairs so that the overhanging wall of one cell wall is matched by a corresponding condition in the adjoining cell. The circular or oval opening in the secondary wall of a bordered pit, called the aperture, leads into the pit cavity formed by the overarching wall. In the wood of pine trees the pit membrane possesses a pad of material called the torus. In nonfunctioning cells, the torus is pulled to one side to block the aperture of a bordered pit, thus preventing the movement of water from one cell to another through the pit region.

Label the parts of a bordered pit in the diagram on the next page.

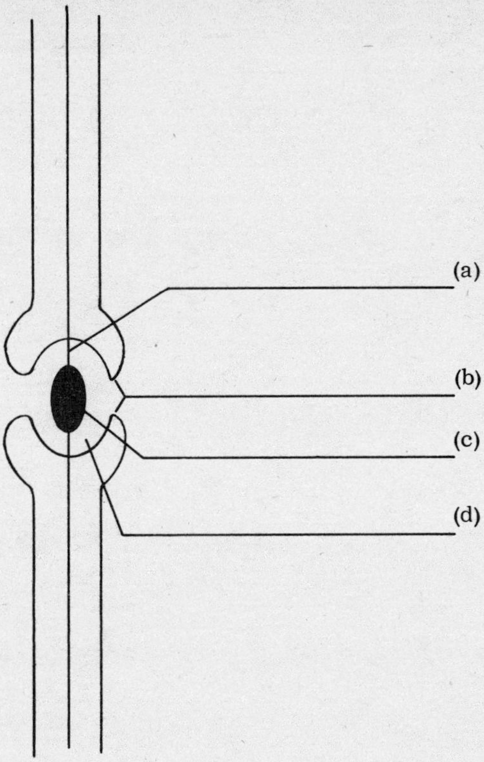

- - - - - - - - - - - - - - - - - -

a. pit membrane; b. aperture; c. torus; d. pit cavity

12. Xylem tissue is composed of two other important components besides tracheids and vessel members. They are xylem fibers and xylem parenchyma.

Xylem fibers function as dead supportive cells with thick lignified secondary walls, and several forms of fibers can be found in xylem tissue. Some resemble tracheids with simple and bordered pits in their walls and with rather large cellular cavities (fiber tracheids), while others show intergrading forms which culminate as fibers with few simple pits and very narrow cell cavities due to excessively thickened walls (libriform fibers). They are similar to the sclerenchyma fibers described in Chapter 1 (see photograph on next page).

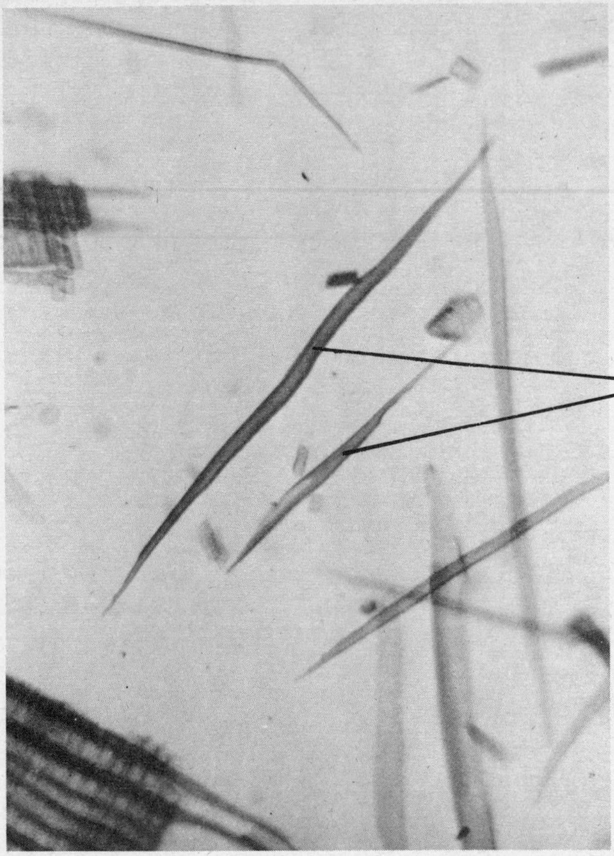

Many plant anatomists consider xylem fibers as modified tracheids whose tiny cell cavities preclude water conduction. Check the statement(s) below which reveal(s) thc similarities between tracheids and xylem fibers.

________ a. Both fibers and tracheids may function as dead cells.

________ b. Libriform fibers have thicker walls in proportion to their diameters than do tracheids.

________ c. Xylem fibers and tracheids are cells that are supportive in nature.

________ d. Both xylem fibers and tracheids possess secondary walls.

- - - - - - - - - - - - - - - - - -

All statements express true relationships between xylem fibers and tracheids.

13. Xylem parenchyma cells, unlike those of simple parenchyma tissue, seldom possess isodiametric dimensions. Their primary walls which vary in thickness are marked with simple pits. Secondary walls may be

formed in xylem parenchyma, and both simple and bordered pits may occur. Xylem parenchyma cells function as typical parenchyma cells in that they possess active protoplasts upon maturity, and they store food as well as other organic materials. Under certain conditions, they reveal their meristematic potential by producing new wound tissue.

In what three ways do xylem parenchyma cells resemble simple parenchyma tissue?

(1) ______________________________

(2) ______________________________

(3) ______________________________

- - - - - - - - - - - - - - - - -

(1) They possess living protoplasts. (2) They may store foods and other organic materials. (3) They are potentially meristematic. (any order)

14. In summary, xylem tissue is composed of four types of cells; they are

______________, ______________, ______________,

and ______________. Of these xylem cell types, which one normally functions as living cells? ______________

- - - - - - - - - - - - - - - - -

tracheids, vessel members, xylem fibers, and xylem parenchyma (any order); xylem parenchyma

15. To further summarize the nature of the xylem units, match the items listed below. Some of the blanks may require more than one answer.

________ a. End walls missing.		1. xylem fibers
________ b. Bordered pits.		2. tracheids
________ c. Possibly represent modified water-conducting cells.		3. xylem parenchyma
________ d. Tapered or pointed end walls.		4. vessel members
________ e. Function primarily as water-conducting cells.		
________ f. Spiral, annular, reticulate wall thickenings.		

- - - - - - - - - - - - - - - - -

a. 4; b. 1, 2, and 4; c. 1; d. 1 and 2; e. 2 and 4; f. 2 and 4

16. As a final summarizing activity on xylem, label the cell types as represented by the photograph below.

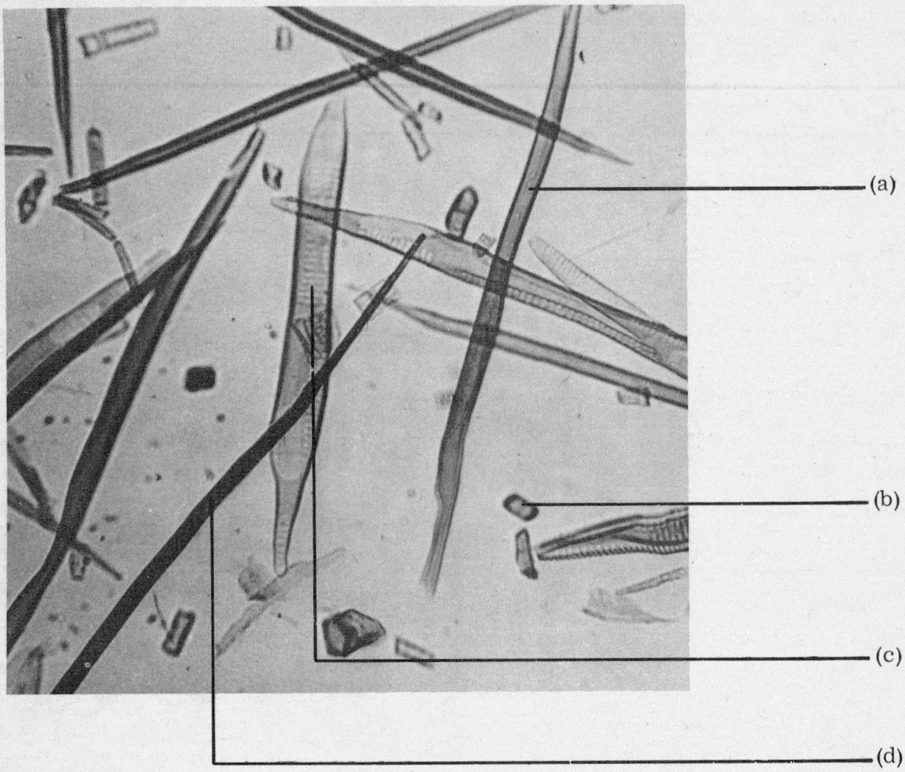

- - - - - - - - - - - - - - - - - -

a. tracheid; b. xylem parenchyma; c. vessel member; d. xylem fiber

17. In the opening frames of this chapter you learned that vascular tissue was composed of two complex tissues, ______________________________ and ______________________________. To review, in what way does complex tissue differ from simple tissue? ______________________________ ______________________________

- - - - - - - - - - - - - - - - - -

xylem, phloem (either order); Complex tissue is composed of more than one kind of cell.

18. Now that you have learned something about complex xylem tissue, let us consider phloem tissue. Most people know that green plants can manufacture their own food, but let us consider a mature tree. Only its leaves

are green, and its branches, trunk, and roots lack the capacity to produce food. Therefore, a food formed by the green aerial portions must be translocated to the other parts of the tree. The phloem tissue serves as the main pathway for food translocation in a vascular plant.

Girdling, which involves the removal of a ring of bark from the trunk of a tree, is often practiced as a means of killing a standing tree. Since girdling severs the phloem tissue which connects the leaves and roots, can you suggest why a tree dies when it is girdled? ________________

____

- - - - - - - - - - - - - - - - -

The removal of a section of phloem prevents food from reaching the roots from the leaves and the roots die from the lack of food. This ultimately results in the death of the entire tree.

19. When a longitudinal portion of phloem tissue is examined under the microscope, many cells appear somewhat tubular and thin-walled. These are the <u>sieve tube members</u> which are laid down in a vertical series by the meristems. Their protoplasts are active while the cells are immature, but become sluggish and visibly inactive as the cells cease growing. Moreover, each <u>nucleus</u> (plural, nuclei), a spherical protoplasmic portion of the protoplast which directs the chemical and hereditary aspects of the living cell, disintegrates. Thus, mature sieve tube members are unique among most living plant cells in that they lack organized nuclei.

    Sieve tube members are to phloem what vessel members are to xylem. What significant difference exists between these two kinds of cells?

____

____

- - - - - - - - - - - - - - - - -

A vessel member is dead upon maturity and lacks a protoplast, while a sieve tube member retains its protoplast minus the nucleus.

20. Careful examination of the cross walls which separate one sieve tube member from its neighboring cells lying above and below it reveals clusters of sievelike perforations (the <u>sieve areas</u>) through which protoplasmic strands of varying diameter link the protoplasts of the sieve tube members in a vertical series. The part of a wall of a sieve tube member which bears one or more sieve areas is commonly called a <u>sieve plate</u>. While sieve plates usually occur on the end walls of a sieve tube member, they may occasionally occur on its side walls as well.

    The tubelike nature of the vertically arranged cells together with their sievelike cross walls, suggest the name <u>sieve tube</u>, which is applied to each series of cells. Thus, sieve tubes composed of sieve tube members are analogous to the vessels and vessel members in xylem.

Label sieve tube, sieve tube members, and perforations in the following diagram.

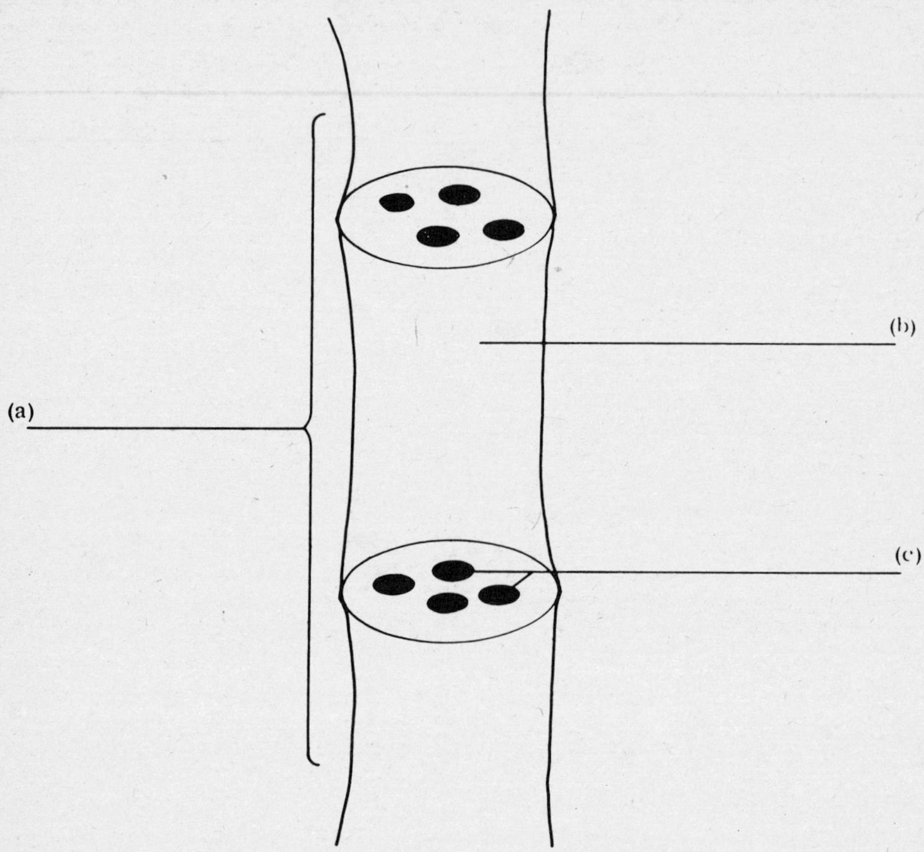

- - - - - - - - - - - - - - - - -

a. sieve tube; b. sieve tube member; c. perforations

21. Sieve tube members are always accompanied by companion cells in flowering plants. Like parenchyma in appearance, with prominent nuclei and thin walls, companion cells show an intimate relationship with sieve tube members in manner and time of origin from the meristems. Since little is known about the mechanism involved in the movement of soluble foods through the phloem tissue, the functional relationship between companion cells and sieve tube members is unclear. The discovery of plasmodesmata linking the sieve tube members with the companion cells also suggests a close physiological relationship between the non-nucleated sieve tube members and the nucleated companion cells.

The protoplasmic content of companion cells differs from that of sieve tube members in what respect? ______________________

- - - - - - - - - - - - - - - - -

Companion cells have nuclei, while mature sieve tube members do not.

22. Phloem tissue, like xylem tissue, also contains parenchyma and sclerenchyma. Phloem parenchyma cells store food and other accumulated products, and, in some instances, they conduct food laterally. Phloem parenchyma will be discussed in greater detail in subsequent chapters dealing with root and stem anatomy.

Sclerenchyma occurs primarily as phloem fibers which function as supportive members of the phloem. Their presence in large amounts often gives commercial value to phloem tissue as a source of cordage, mentioned previously. In addition to fibers, sclerenchyma may occur in older accumulations of phloem as sclereids.

The linden tree is often called basswood, a corruption of the name bastwood. Bast was a term formerly used for phloem, and wood as used in this sense implies a hardness or toughness of the bast. What cells in phloem tissue are responsible for giving the linden tree its name basswood? ______________________________

- - - - - - - - - - - - - - - - -

In linden, phloem fibers occur in such large amounts that the bark containing the phloem is quite fibrous.

23. Phloem is classed as a complex tissue. Four components of phloem are

____________________, ____________________, ____________________, and ____________________.

- - - - - - - - - - - - - - - - -

sieve tube members, companion cells, phloem fibers, and phloem parenchyma (any order)

24. To summarize the structure and the nature of phloem tissue, write the correct name of the phloem component in the space provided before each of the following statements.

____________________ a. Cells often functioning in phloem as dead cells.

____________________ b. Cells intimately associated with the main food-conducting cells in phloem.

____________________ c. Cells functioning in storage and lateral conduction of food.

____________________ d. A living cell without nuclei upon maturity.

____________________ e. Cells having large perforations and strands of protoplasm linking two adjacent cells.

(continued)

________________ f. A vertical series of food-conducting cells.

- - - - - - - - - - - - - - - - - -

a. phloem fibers; b. companion cells; c. phloem parenchyma; d. sieve tube members; e. sieve tube members; f. sieve tube

25. Compare and contrast xylem and phloem with respect to their functions.

____

____

____

- - - - - - - - - - - - - - - - - -

Xylem and phloem are both complex tissues which function in conduction. Xylem conducts water and dissolved minerals from the roots to the stem and leaves. Phloem translocates food manufactured in the green leaves to the stem and roots.

26. Compare and contrast xylem and phloem by indicating whether each of the following statements is:

    1. True of both xylem and phloem
    2. True of xylem but not of phloem
    3. True of phloem but not of xylem
    4. True of neither xylem nor phloem

________ a. Complex tissue.

________ b. Translocates food from aerial plant parts to the roots.

________ c. At maturity consists primarily of dead cells.

________ d. At maturity consists entirely of dead cells.

________ e. Includes cellular components called tracheids and vessels.

________ f. Includes living non-nucleated cells.

________ g. Includes both parenchyma and sclerenchyma cells.

________ h. Conductive or vascular tissue.

________ i. Includes a kind of cell called a companion cell.

________ j. Translocates water and dissolved nutrients from roots to leaves.

________ k. Includes cells called fibers.

- - - - - - - - - - - - - - - - - -

a. 1; b. 3; c. 2; d. 4; e. 2; f. 3; g. 1; h. 1; i. 3; j. 2; k. 1

SELF-TEST

Before you proceed, take this short Self-Test to help you reinforce what you learned in this chapter. Although a few of the questions relate to Chapter 1, most are to be answered in the light of information in Chapter 2 of this book. Answers are given at the end of this quiz.

1. Simple plant tissues differ from complex plant tissues in what major way?

 ____

 ____

2. Give at least two examples of simple plant tissues. ________________

 ____

3. Give two examples of complex plant tissues. ____________________

 ____

4. Compare and contrast the two complex plant tissues in terms of their function. __________________________________

 ____

 ____

5. What are two special features that are especially characteristic of the cell walls of both tracheids and vessels? ____________________

 ____

6. Listed below, numbered in alphabetical order, are the various kinds of cells found in xylem and phloem. Match the cell type with its corresponding description. Some of the blanks may require more than one answer.

1. companion cells	5. tracheids
2. phloem fibers	6. vessel members
3. phloem parenchyma	7. xylem fibers
4. sieve tube members	8. xylem parenchyma

________ a. Narrow, elongated, water-conducting cells with tapered, somewhat pointed end walls; dead at maturity.

________ b. Living tubular, thin-walled cells, lacking nuclei at maturity, and having cross walls perforated with protoplasmic strands.

________ c. Bast fibers; often dead cells; especially abundant in the bark of linden trees.

________ d. Dead cells with thick, lignified secondary walls, but with end walls lacking at maturity.

(continued)

_________ e. Thin-walled, parenchyma-like cells with pronounced nuclei; located adjacent to sieve tube members in flowering plants; precise function uncertain.

_________ f. Living xylem cells, rarely found to be isodiametric; function in food storage; potentially meristematic.

_________ g. Living phloem cells; store food; may conduct food laterally.

_________ h. Water-conducting cells characterized by having secondary walls that have scalariform, reticulate, or pitted thickening patterns.

_________ i. Dead supportive phloem cells with thick lignified secondary walls.

7. Name two simple tissue cell types which may also be found in complex tissues. ____

8. Identify each of the cell types shown in the diagrams on the next page, selecting your answers from among the numbered choices below.

   1. companion cells
   2. phloem parenchyma cell
   3. pitted vessel
   4. reticulate vessel
   5. scalariform vessel
   6. sieve tube
   7. tracheid
   8. tracheid with spiral thickening
   9. xylem parenchyma cell

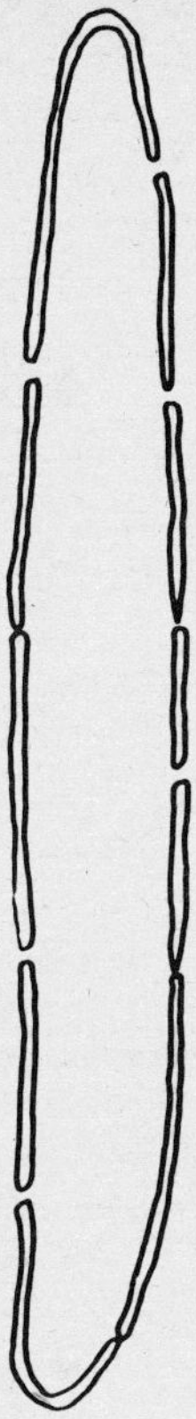

(a) ____________________

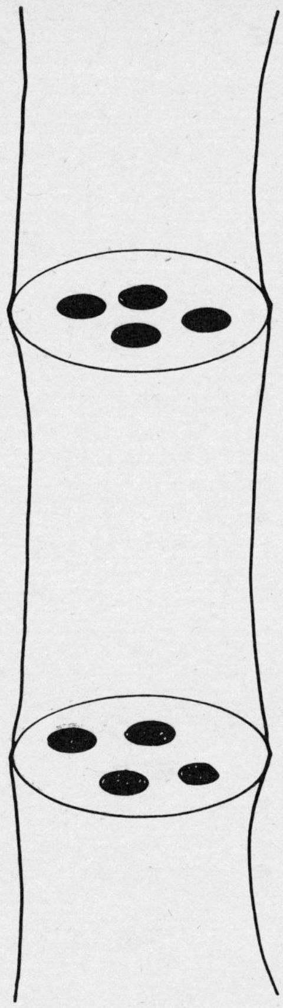

(b) ____________________

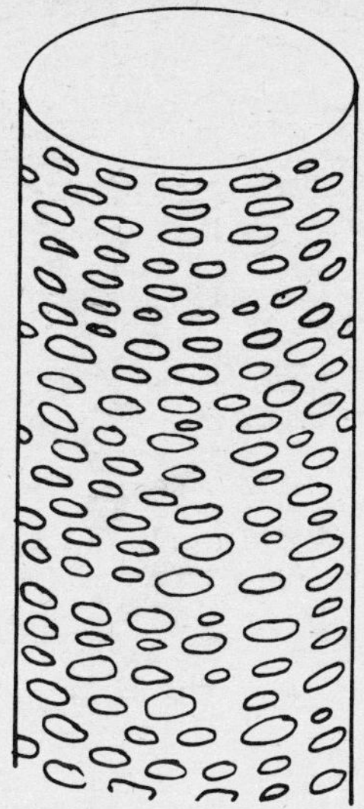

(c) ____________________

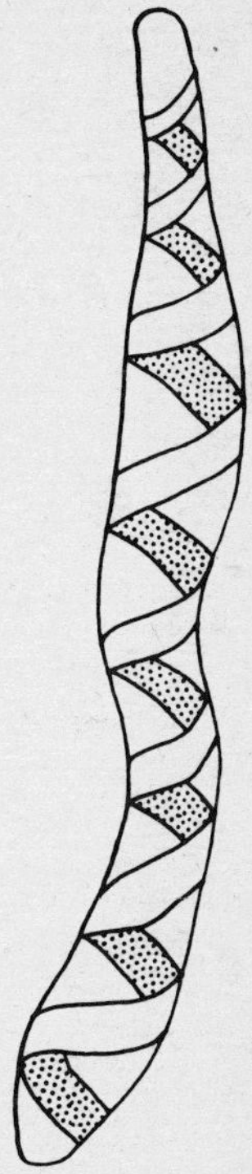

(d) ____________________

9. Label the parts of the bordered pit in the diagram on the next page.

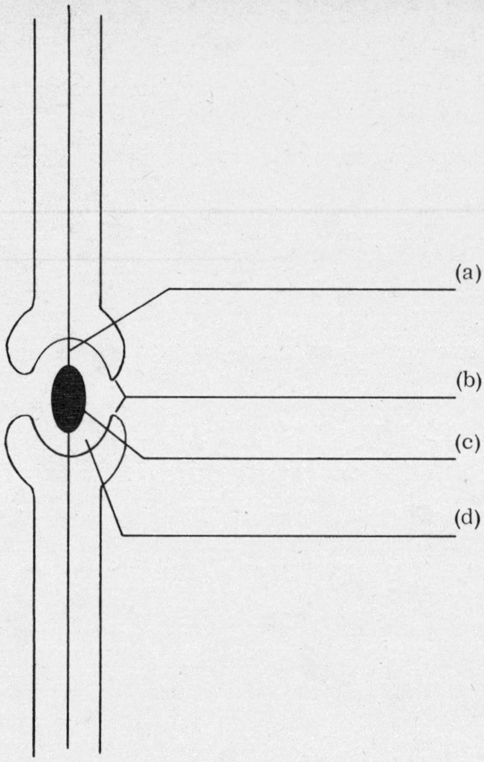

10. Pits and pit structure have been emphasized in this chapter because they are useful in helping the plant anatomist identify and characterize certain kinds of cells in the plant. They are also interesting and may be of importance to the plant itself because they are sites of protoplasmic bridges that connect adjacent cells. What name is given to the living bridges of protoplasm that connect adjacent cells at the sites of the pits? ________

____

Answers

Compare your answers to the Self-Test with the correct answers given below. If all of your answers are correct, you are ready to go on to the next chapter. If you missed any questions, review the frames indicated in parentheses following the answer before you go on.

1. Simple plant tissues are composed of but one kind of cell, while complex tissues are made up of several different kinds of cells. (frame 1)
2. Parenchyma, collenchyma, sclerenchyma, epidermis, and cork are all examples of simple tissues discussed in Chapter 1. (Chapter 1, frames 24, 25, 29, 31, 37)

3. xylem and phloem (frame 2)

4. Both xylem and phloem are conductive tissues, but xylem translocates water and dissolved materials from the roots to the aerial parts of the plant while phloem conducts food manufactured in the green leaves down the stem and to the roots. (frames 2, 3, 18)

5. Characteristic thickenings in the secondary wall of tracheids and vessels; the presence of noticeable pits in the cell walls. (frames 7-9)

6. a. 5 (frame 4)
   b. 4 (frames 19-20)
   c. 2 (frame 22)
   d. 6 (frame 5)
   e. 1 (frame 21)
   f. 8 (frame 13)
   g. 3 (frame 22)
   h. 5 and 6 (frame 8)
   i. 2 (frames 22-24)

7. parenchyma and sclerenchyma (frames 12, 13, 22)

8. a. 7 (frame 4)
   b. 6 (frames 19-20)
   c. 3 (frame 8)
   d. 8 (frame 7)

9. a. pit membrane; b. aperture; c. torus; d. pit cavity (frame 11)

10. plasmodesmata (Chapter 1, frame 7)

CHAPTER THREE

Embryos and Seedlings

In the two previous chapters you learned of the nature of the major tissues which comprise a flowering plant. We are ready now to consider how these tissues originate and how they become arranged to form the three major organs of the plant; namely, the roots, stems, and leaves. We may start by studying the inception of the plant embryo and its early development.

1. As you perhaps already know, a flowering plant reproduces chiefly by the formation of seeds, whereby a young plant, the embryo, comes into existence while protected and nourished by enclosing tissues.

   The fully developed embryo remains within the seed until favorable environmental conditions induce it to resume growth (germinate) and push through the soil as a young plant called a seedling.

   A plant begins its existence as an embryo enclosed by tissues of the ____________________. Under proper conditions, the embryo develops into a(n) ____________________.

   - - - - - - - - - - - - - - - - -

   seed; seedling

2. Suppose you examined microscopically very thin sections of a developing seed. In doing so, you may discover an egg cell which has just been fertilized by a sperm cell. The fertilized egg marks the beginning of a new plant. (For more detailed discussion of these matters, see the Self-Teaching Guide Plant Life Cycles, by Mertens and Stevenson, Wiley, 1975.) Further examination of sections of progressively older seeds reveals that the fertilized egg undergoes a series of cell divisions to establish a cylindrical, multicellular proembryo. Acceleration of the rate of cell division at one end of the proembryo results in the appearance of a spherical embryo, to which is attached the remainder of the filamentous portion which was the proembryo but is now called the suspensor (see diagram on next page).

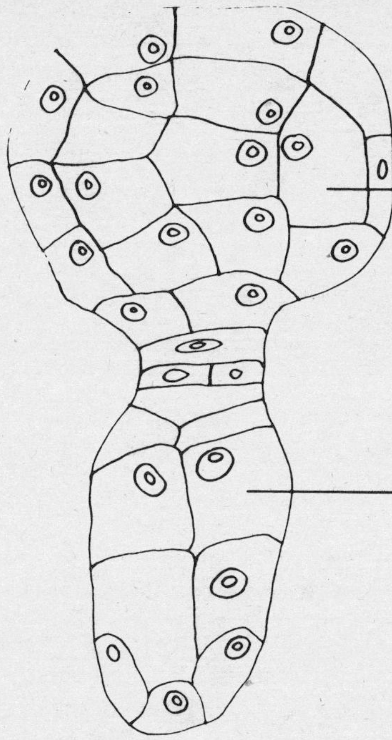

A proembryo develops into a filamentous ______________________

and a spherical ________________________.

- - - - - - - - - - - - - - - - - -

suspensor; embryo

3. Rapid changes now ensue in the embryo as successive cell divisions and cell enlargement cause the embryo to flatten and lose its spherical shape. Lateral outgrowths change the once spherical embryo into a heart-shaped structure.

Listed below are some stages that occur in the formation of a seed embryo. Put these lettered items in correct chronological order to indicate the sequence in which they appear.

(A) suspensor, (B) proembryo, (C) fertilized egg, (D) embryo, (E) egg

____

- - - - - - - - - - - - - - - - - -

E—C—B—A and D (latter two in either order)

4. An early differentiation occurs among the meristematic cells of the young embryo which will eventually give rise to the primary tissues and apical meristems of the young seedling. A superficial layer of cells develops

into a distinct protoderm which will give rise to the epidermis of the seedling. Other meristematic cells situated more internally begin to elongate to form strands of procambium, the forerunner of the vascular tissue. The remainder of the cells lying inside the protoderm and interrupted by strands of procambium mark the ground meristem. It will form parenchyma, collenchyma, and sclerenchyma, or the so-called ground tissues (see photograph).

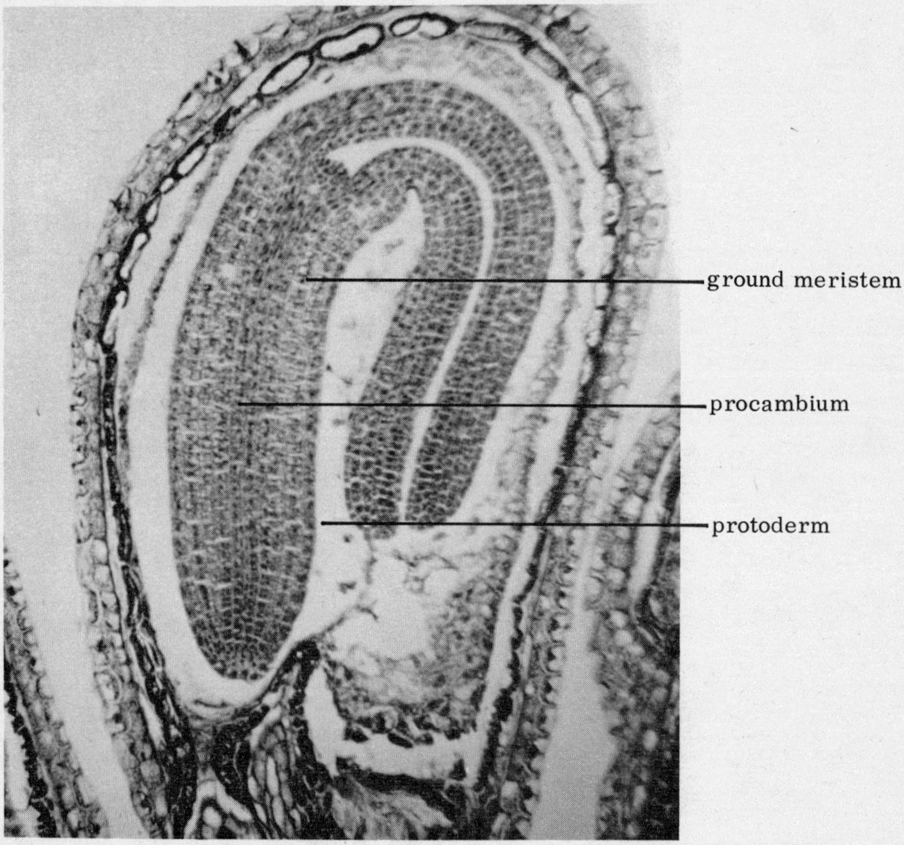

Match the words and definitions listed below.

________	a. Forms epidermal tissue.	1.	ground meristem
________	b. Differentiates into xylem and phloem.	2.	procambium
________	c. Chief source of parenchyma.	3.	protoderm
________	d. Origin of vascular tissue.		
________	e. Source of ground tissue.		

- - - - - - - - - - - - - - - - - -

a. 3; b. 2; c. 1; d. 2; e. 1

5. Subsequent differentiation results in the formation of a mature embryo consisting essentially of an elongated axis to which are attached two leaf-like appendages, the cotyledons (cot´-uh-lee-dons).

The portion of the axis of the embryo below the point of cotyledon attachment is appropriately called the hypocotyl (literally, "below the cotyledons"). Because the lowermost portion of the hypocotyl forms a different part of the seedling than does the hypocotyl proper, the lowermost end of the axis is often called the radicle.

The portion of the embryo axis lying above the point of cotyledon attachment is often a mound of meristematic tissue marking the location of the apical meristem. In many embryos, however, primary tissue may develop from the apical meristem to form an epicotyl (literally, "above the cotyledons"), at the tip of which is the apical meristem (see diagram).

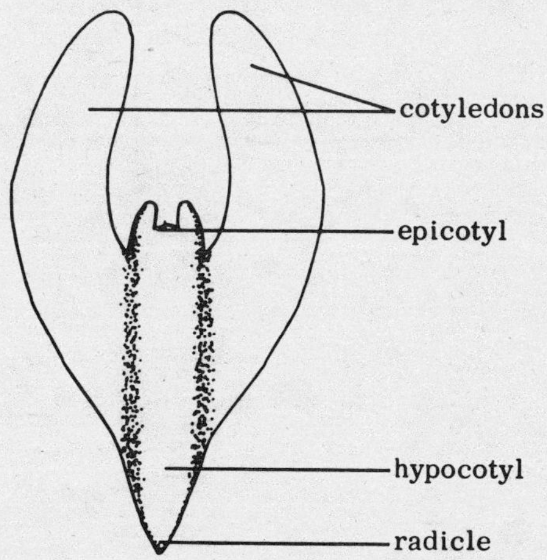

In summary, the plant embryo is composed of the following four regions: ____

____.

- - - - - - - - - - - - - - - - - -

cotyledons, hypocotyl, radicle, and epicotyl (any order)

6. Not all embryos of flowering plants produce two cotyledons in the manner just described. Many embryos, such as those of wheat, illustrated at the top of the next page, form a single cotyledon, and a hypocotyl is not present.

The portion of the wheat embryo labeled epiblast in the photograph has no apparent function, and some anatomists interpret it as a rudimentary cotyledon.

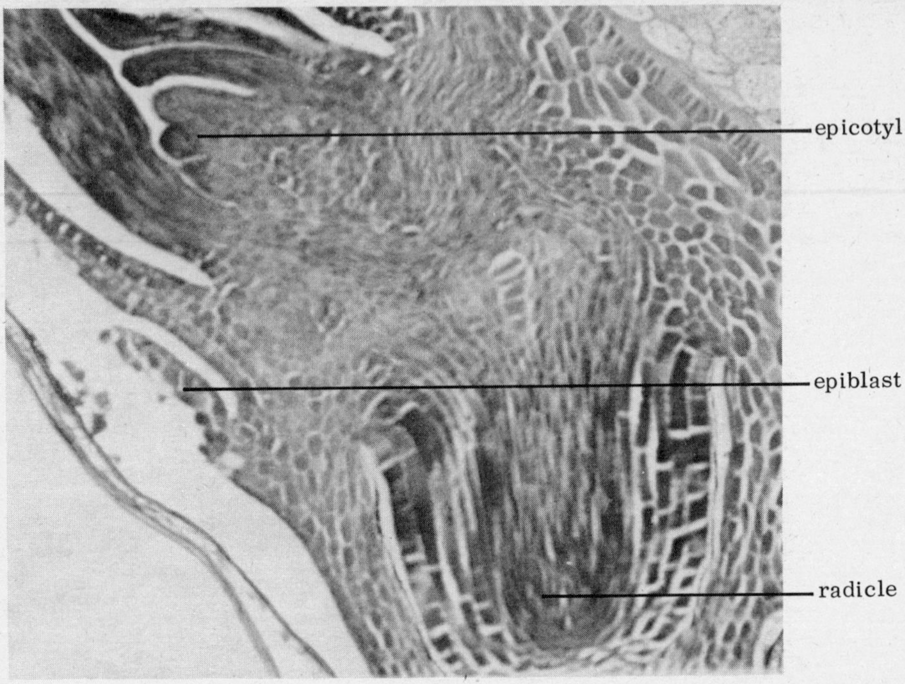

The corn embryo, like wheat, has but a single cotyledon and is often used to demonstrate the occurrence of one cotyledon. Illustrated below is the corn embryo seen in surface view. Since the embryos of wheat and corn are quite similar you should be able to label the parts of the corn embryo.

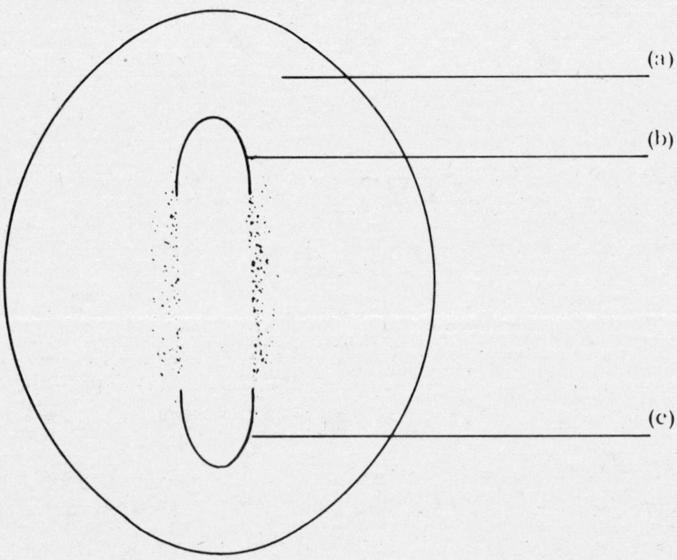

- - - - - - - - - - - - - - - - -

a. cotyledon; b. epicotyl; c. radicle

7. The fact that some flowering plant embryos have two cotyledons and others have one may not, in itself, seem important, but this difference is only one of many which divide the flowering plants into two major groups. Most botanists believe that flowering plants developed along two evolutionary lines since the time of their appearance on earth. These two groups of plants are called dicotyledons and monocotyledons (or briefly, dicots and monocots) based on the number of cotyledons present in the embryos.

Therefore, the embryo depicted in frame 5 of this chapter is a

____________________, while the embryos of corn and wheat

are ____________________.

\- - - - - - - - - - - - - - - - - -

dicot (dicotyledon); monocots (monocotyledons)

8. You should not infer that the number of cotyledons possessed by the embryos of the individuals in the two evolutionary lines is of greater importance than the other differences which separate them. Throughout the remainder of this book your attention will be called repeatedly to the many anatomical differences between monocots and dicots.

The embryo of squash has two cotyledons. Therefore, it is classed

as a ____________________. Would you expect corn and squash plants to have anatomically similar roots, stems, and leaves? Explain.

____

\- - - - - - - - - - - - - - - - - -

dicot; No, since corn is a monocot and the squash plant is a dicot.

9. Suppose you soak a lima bean seed in water to soften the seed coat so that it can be easily removed. You now find an embryo whose two large cotyledons readily separate. The axis of the bean embryo lies between the cotyledons as illustrated below.

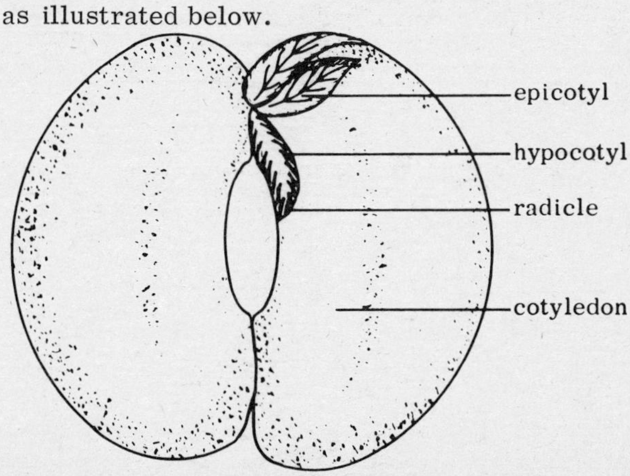

The bean embryo contains much stored food which will be used by the young seedling as it develops from the embryo. Can you guess by referring to the drawing in what part of the embryo the food is stored?

____

- - - - - - - - - - - - - - - - -

The greatly enlarged cotyledons contain the food, which we utilize in our own bodies when we eat beans.

10. The bean embryo is capable of producing a seedling once its dormancy is broken. If an intact bean seed is planted in warm damp soil, germination of the seed begins within a few hours. What happens first? If you have watched seeds germinate you may already know that the root is the first organ of the seedling to break through the seed coat. It arises from the radicle of the embryo.

    Can the terms radicle and root be used interchangeably? Explain.

____

____

- - - - - - - - - - - - - - - - -

No, because the radicle is an embryo structure which gives rise to an organ of the seedling, the root.

11. If you continue to follow the sequence of germination of the bean embryo, you will observe that the hypocotyl greatly elongates and pushes up through the surface of the soil in an arched manner. Continued elongation of the hypocotyl above the soil pulls the cotyledons and epicotyl above ground. As the hypocotyl straightens vertically, the cotyledons, now in an aerial position, separate to expose the epicotyl (see illustration on next page).

    In many plants, such as peas, the cotyledons remain below ground.

    Can you guess why they remain in that position? ____________________

____

- - - - - - - - - - - - - - - - -

There is nothing to bring them above ground because the hypocotyl does not elongate.

12. As the illustration depicting bean germination indicates, the hypocotyl forms the stem portion between the ground line and the cotyledons. The remainder of the stem and the leaves develop from the epicotyl. The cotyledons provide stored food to the young bean plant until its leaves can manufacture food.

The stem of the bean seedling is formed from what two parts of the embryo? ______________________________

- - - - - - - - - - - - - - - - -

hypocotyl and epicotyl

13. As you learned in frame 11, some hypocotyls do not function. Consequently, the stems in these plants must arise from ____________________.

- - - - - - - - - - - - - - - - -

epicotyls

14. The sequence of events in the germination of a bean seed is basically the same for seeds of almost all other plants. Variations in germination arise chiefly from the behavior of the hypocotyls, as already discussed, and from the activity of the cotyledons. Cotyledons in all plants provide nourishment for the seedlings, and this may be accomplished in several

ways: (1) cotyledons may act as storage organs as in the bean and pea, or (2) they may assume the function of leaves temporarily and produce food from sunlight as in squash.

A botany student notes that a young pumpkin plant has two kinds of leaves—those that he recognizes as characteristic of pumpkin plants plus a single pair of rounded, flat green leaves that are located near the soil surface. What might these rounded, flat green leaves be? What function do they have? ______________________________

____

- - - - - - - - - - - - - - - - -

The rounded, flat green leaves are the cotyledons which, because they are green, can carry on the food manufacturing process.

15. Cotyledons may nourish young seedlings in yet a third way: they may absorb food stored in the seed and transfer it to the seedling.

The embryo in a grain of corn lies with its single cotyledon in close contact with an abundant supply of food stored outside of the embryo itself. As the corn grain germinates, the root emerges first, followed by the stem and a true leaf differentiated from the epicotyl. Since the hypocotyl does not elongate, the cotyledon and the stored food in the corn grain remain below ground (see diagram).

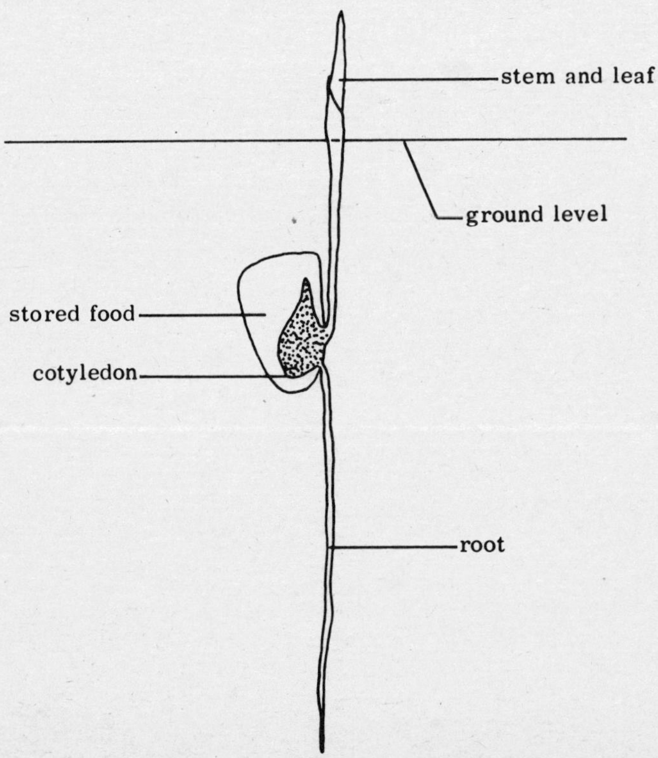

Although the true leaf has emerged above ground, it does not manufacture much food at this time, and the future growth of the young corn seedling is dependent upon the stored food supply. Can you suggest how the seedling is able to utilize this food? ______________________

____

- - - - - - - - - - - - - - - - -

The cotyledon absorbs the food and transfers it to the growing seedling.

16. Regardless of the manner in which cotyledons function as organs of nourishment for the seedling, they are transitory structures. Eventually cotyledons become detached from the actively growing seedling. Can you guess why it is possible for the seedling to survive the loss of its cotyledons? ______________________________________

____

- - - - - - - - - - - - - - - - -

By the time the cotyledons are lost the seedling has developed its first true leaves containing the green pigment that allows the seedling to manufacture its own food. It is therefore no longer dependent on the cotyledons for nourishment.

17. Summarize your understanding of their function by listing the three ways in which cotyledons may affect the nourishment of the embryo and young seedling. ______________________________________

____

____

- - - - - - - - - - - - - - - - -

Cotyledons may (1) store food, (2) in the presence of sunlight manufacture food for the young seedling, and (3) absorb and transfer to the seedling food stored in the seed.

18. To further summarize what you have learned about the plant embryo and its development into a seedling, complete the following matching exercise. Some of the blanks may require more than one answer.

________ a. Portion of the embryo that develops into the root.

________ b. Portions of the embryo that may develop into the stem.

1. epicotyl
2. radicle
3. cotyledon
4. hypocotyl

(continued)

________ c. Portion of the embryo that gives rise to the true leaves.

________ d. Portion of the embryo that stores, manufactures, or absorbs food.

________ e. Portion of the bean embryo that first emerges above the soil.

- - - - - - - - - - - - - - - - -

a. 2; b. 1 and 4; c. 1; d. 3; e. 4

SELF-TEST

Before you proceed, take this short Self-Test to help you reinforce what you have learned in this chapter. All questions are to be answered in the light of information in Chapter 3 of this book. Answers are given at the end of this quiz.

1. Place the following lettered items in correct chronological order to indicate the successive stages in the formation of a young seedling.

   (A) suspensor + embryo, (B) egg, (C) root, stem, and leaves of seedling, (D) proembryo, (E) hypocotyl + epicotyl + cotyledons of mature embryo, (F) fertilized egg

   ____

2. Match the correct numbered items with the corresponding lettered items.

   ________ a. Origin of root.

   ________ b. Stem = ____ + ____.

   ________ c. Origin of epidermis.

   ________ d. Forerunner of vascular tissue.

   ________ e. Organs of food storage, production, and absorption.

   ________ f. Forms parenchyma and collenchyma.

   ________ g. Gives rise to xylem and phloem.

   1. cotyledons
   2. epicotyl
   3. ground meristem
   4. hypocotyl
   5. procambium
   6. protoderm
   7. radicle

3. So far we have stressed one major difference between monocots and dicots. What is this major difference? ______________________________

   ____

4. Give at least one example of a monocot. ______________________

   Of a dicot. ______________________

5. In the following diagram of a bean seed, identify the labeled parts, selecting your answers from among the numbered items listed to the right.

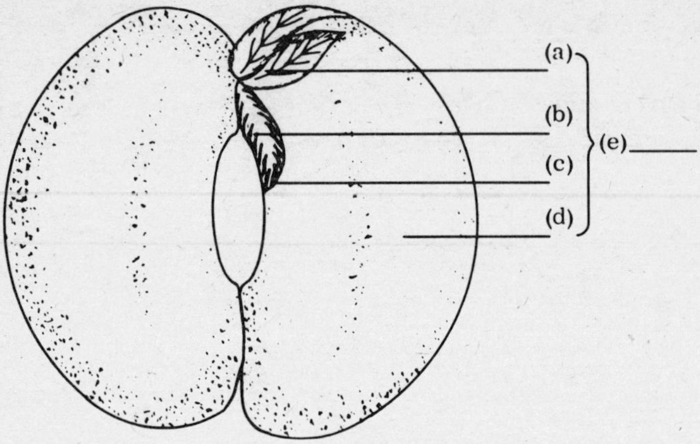

1. cotyledon
2. embryo
3. epiblast
4. epicotyl
5. ground meristem
6. hypocotyl
7. procambium
8. protoderm
9. radicle

6. Refer back to the drawing of a bean seed in question 5 and answer the following questions.

   a. Part a gives rise to what organ(s) of the seedling? ______________

   ____

   b. Part b gives rise to what organ(s) of the seedling? ______________

   ____

   c. What is/are the function(s) of part d? ______________________

   ____

   d. Which part (a, b, c, or d) first appears above the surface of the soil when the bean seed germinates? ______________

   e. Part c gives rise to what organ(s) of the seedling? ______________

   ____

Answers

Compare your answers to the Self-Test with the correct answers given below. If all of your answers are correct, you are ready to go on to the next chapter. If you missed any questions, review the frames indicated in parentheses following the answer before you go on.

1. B—F—D—A—E—C (frames 1-13)

2. a. 7 (frame 10)
   b. 2 and 4 (frames 11 and 12)
   c. 6 (frame 4)
   d. 5 (frame 4)
   e. 1 (frames 14 and 15)
   f. 3 (frame 4)
   g. 5 (frame 4)

3. Monocots have one seed leaf while dicots have two. (frame 7)

4. Monocot: wheat, corn (frames 7 and 8)
   Dicot: bean, squash, pumpkin (frames 8, 9, 14)

5. a. 4; b. 6; c. 9; d. 1; e. 2 (frame 9)

6. a. leaves and stem (frames 12 and 13)
   b. stem (frames 11 and 12)
   c. store food, manufacture food, and in some plants, absorb food stored in the seed (frames 14 and 15)
   d. b, the hypocotyl (frame 11)
   e. the root (frame 10)

CHAPTER FOUR
The Root

1. You learned in the previous chapter that the root arises from the activity of the radicle portion of the embryo. The early differentiation of the root is a distinct advantage to the young seedling for it is an important organ with diverse activities. A root system not only anchors the plant, but it absorbs water and nutrients from the soil. During the growing season the roots conduct water to the organs of the plant above ground. Food, such as sugar and starch, is manufactured in the leaves and may be translocated to the roots. Thus, roots may also perform the important function of food storage, especially during the winter.

In early spring, maple trees are often tapped to collect the sugary sap which rises in the stem from below ground. What two functions of the roots of maple trees make this possible? ______________________

____

- - - - - - - - - - - - - - - - -

Food is stored in the roots of the maple tree during the winter months. Water and food are conducted upward by the roots in the spring.

2. If roots originated solely from the embryo, a plant would possess only one root. However, anyone who has removed a potted plant from its container knows that this is not true. Roots may arise from two additional sources: (1) they may form from other roots, and (2) they may arise from other organs of the plant besides the roots.

A common method of propagating a plant is by cuttings, in which the top portion of the stem containing the leaves is cut off and the cut end is placed in damp sand or water. Eventually a mass of roots grows from the cutting. From which of the three sources discussed in this frame do these roots arise? ______________________________

- - - - - - - - - - - - - - - - -

They originate from the stems.

3. The first root formed from the radicle is considered to be the primary root. Almost immediately after its formation the primary root produces numerous branch or secondary roots. Secondary roots, in turn, give rise to tertiary roots and so on until an elaborate root system is formed. If the primary root and its branching members remain uniform in diameter, the system is referred to as a fibrous root system.

A fibrous root system is composed of equal sized roots. What names are given to the first three sets of roots in relation to the order in which they develop? ______________________________

- - - - - - - - - - - - - - - - -

primary, secondary, and tertiary roots

4. In most dicots the primary root becomes much larger than its branch roots and forms a taproot. Such a root usually grows to a great length. Sometimes taproots become greatly enlarged with stored food. A taproot plus its branch roots forms a taproot system.

We eat the roots of carrots, parsnips, turnips, and radishes. What type of root system do these plants possess? ____________________

- - - - - - - - - - - - - - - - -

They all possess a taproot system in which food is stored in the taproot.

5. Most monocots form a fibrous root system in which a mass of roots produced from the lower portion of the stem replaces the short-lived taproot. Roots arising from sources other than roots or the embryo are called adventitious roots.

Grasses are monocots and most grasses form a fibrous root system. What inference can you make regarding the probable origin of grass roots?

____

- - - - - - - - - - - - - - - - -

In all probability they are adventitious roots which arise from the stem.

6. Taproots and fibrous roots both bestow certain advantages on those plants which possess them. Some taproots grow downward to great depths to obtain soil water which is not available to more shallowly rooted plants in times of drought. Fibrous roots, however, enable a plant to retain soil around its roots, thus preventing excessive soil erosion.

The statement was made previously that most dicots possess taproots and that most monocots form fibrous roots. Plants of which group would be more successful in holding soil in place on a hillside? Why? (Write your answer on the next page.)

____

____

- - - - - - - - - - - - - - - - -

Monocots with fibrous roots would be better able to retain the soil where erosion, if allowed to occur, would uproot the plants.

7. At this point review the nature of roots by matching the words and phrases below.

________ a. Adventitious roots are associated with this root system.		1. adventitious roots
________ b. The lawn enthusiast trying to rid his lawn of dandelions soon discovers that the dandelion is difficult to uproot because it has a long, slender ____ which penetrates deeply into the soil.		2. taproot
________ c. Conservation minded farmers wishing to prevent soil erosion on a hillside would be wise to grow plants having what kind of root system?		3. fibrous root
________ d. Corn plants are distinguished by having prop roots that develop from the stem and support the upright growth of the plant. Technically such roots are known as ____ because of their origin from the stem.		

- - - - - - - - - - - - - - - - -

a. 3; b. 2; c. 3; d. 1

8. Let us now consider the development of the primary root from the apical meristem. Suppose you placed some radish seeds on a damp paper towel in a closed container. In a few days you would find that the primary root had formed as illustrated on the next page. Although the root possesses several regions of activity, these regions are not easily seen when the root is viewed externally. They have been identified in the drawing for subsequent discussion.

Is the root shown in the illustration most likely to become an adventitious root or a taproot? Explain. ____________________

____

- - - - - - - - - - - - - - - - -

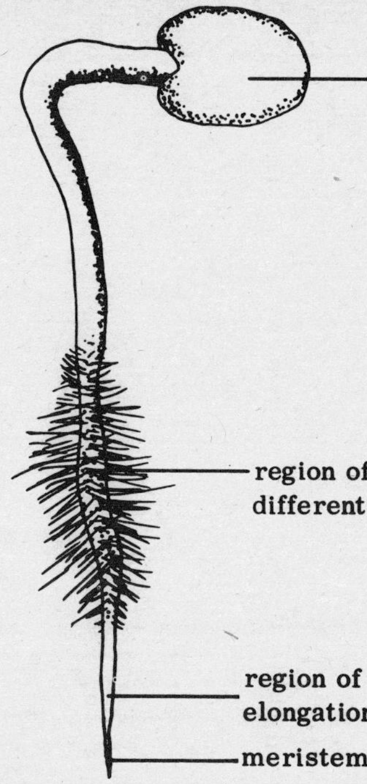

The root shown in the illustration will develop as a taproot because it has arisen from the radicle portion of the embryo. Adventitious roots by way of contrast arise from sources other than the radicle or other roots.

9. The meristematic region occupies only a millimeter or so of the tip of the root. If a thin longitudinal section of the root tip is examined microscopically, the meristem is seen to be composed of actively dividing cells. The accumulation of new cells which are proliferated from the meristem will eventually form the first or primary root tissues, except for a few parenchyma cells which accumulate at the extreme tip of the root. These short-lived parenchyma cells cover the meristematic region with a thimble-like protective layer of cells called the rootcap (see illustration on next page).

    The cells of the rootcap are constantly replaced by the dividing meristem as the older cells of the cap are sloughed off.

    Although described as an apical meristem, the root tip meristem does not, in fact, occur at the extreme tip of the root. Why is this so?

____

- - - - - - - - - - - - - - - - -

The rootcap occurs at the extreme tip of the root and the meristem is literally buried in it.

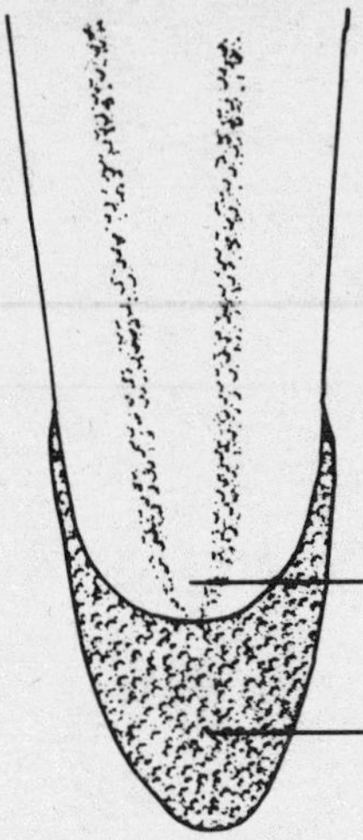

10. The meristematic cells lying inside the rootcap show an early differentiation into protoderm, ground meristem, and procambium, which are the precursors of the primary tissues of the root (Chapter 3, frame 4).
    What kind of tissue will be formed by each of the following?

    a. procambium ______________________________

    b. protoderm ______________________________

    c. ground meristem ______________________________

    - - - - - - - - - - - - - - - - - -

    a. vascular tissue; b. epidermis; c. ground tissue (parenchyma, sclerenchyma, collenchyma)

11. The cells formed by the meristematic region begin to increase in length as growth begins, while the protoderm, ground meristem, and procambium regions become even more pronounced. This marks the <u>region of elongation</u>.
    The elongating cells thrust the meristematic region and rootcap through the soil. Therefore, the three regions of the root tip which literally move through the soil are the ______________________________ ______________________________. Together, they constitute only a few millimeters of the root tip.

    - - - - - - - - - - - - - - - - - -

    rootcap, meristematic region, region of elongation (any order)

12. The role of the rootcap becomes clear when you learn that the meristematic region is pushed through the soil by the region of elongation. The somewhat mucilaginous or slippery nature of the cells together with the physical barrier which the rootcap cells impose between the meristem and the soil afford protection for the delicate meristematic cells.

Can you explain why the mucilaginous character of rootcap cells may be an advantage to the elongating root? ______________________________

____

- - - - - - - - - - - - - - - - -

The rootcap cells may lubricate the pathway so that the root can slip more easily through the soil.

13. The young root cells, once they cease to elongate, begin to differentiate into mature primary tissues. They form another region of the root which lies just above the region of elongation (see diagram). This region of differentiation can best be studied microscopically, but the region is plainly visible to the unaided eye because of the manner in which the epidermal cells differentiate from the protoderm. Many of the epidermal cells form lateral tubular extensions called root hairs which are many times longer than the cells which produce them. You should keep in mind that a root hair is but a part of a single epidermal cell and it is visible to the naked eye due to its great length. Because the region of differentiation is enveloped by root hairs, the term root hair region is often used interchangeably.

Label the indicated portions of the drawing below.

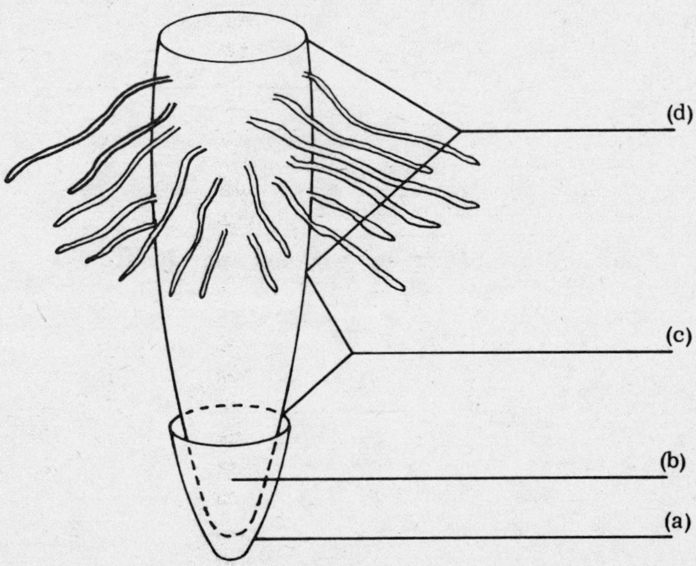

- - - - - - - - - - - - - - - - -

a. rootcap; b. meristematic region; c. region of elongation; d. region of differentiation or root hair region

14. The various tissues which comprise the root may be studied by a microscopic examination of a thin cross section made in the region of differentiation or root hair region. The root in cross section is divided naturally into three general regions corresponding to the tissues differentiated by the protoderm, ground meristem, and procambium. These regions are the epidermis, the cortex, and the stele, as illustrated by the following diagram.

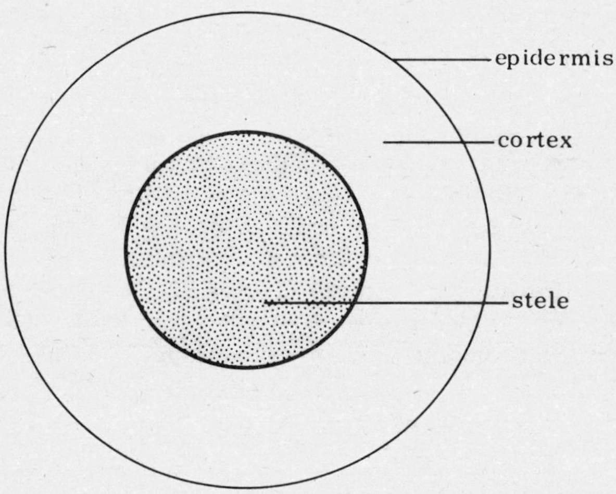

From what you have learned previously, you can expect that the protoderm forms the ________________ region of the root which is classed as (primary/secondary) ________________ tissue.

- - - - - - - - - - - - - - - - - -

epidermal; primary, because it is formed by an apical meristem

15. Match the numbered root regions with the appropriate description.

______ a. Region where cell size is rapidly increasing.		1. meristematic region
______ b. Cells that protect the growing root tip.		2. region of differentiation
______ c. Region where epidermis, cortex, and stele first become evident.		3. region of elongation
		4. rootcap

(continued)

________ d. Region where cell number is rapidly increasing.

________ e. Root hairs externally distinguish this part of the root.

- - - - - - - - - - - - - - - - -

a. 3; b. 4; c. 2; d. 1; e. 2

16. The epidermis of a root is one cell thick. These typical epidermal cells function as living cells upon maturity. You learned that epidermal cells often secrete waxy material called cutin which forms a covering called the cuticle. Since one of the functions of the root is to absorb water and dissolved minerals, you may be surprised to learn that a thin cuticle may be formed over the root epidermis.

Further examination of the root epidermis reveals the absence of stomata and the presence of root hairs arising from many of the epidermal cells. The delicate root hairs increase the surface area of the root to such an extent that most of the soil water absorbed by the roots enters through the root hair region.

The leaves of many plants wilt for a few days after the plants have been repotted. Can you explain why this may be so? ________________

____

- - - - - - - - - - - - - - - - -

Many of the root hairs may have been destroyed during transplanting and the plant is unable to replace the water which is lost from normal evaporation from the leaves.

17. The cortex consists of a wide band of tissues which separates the epidermis from the internal stele. Arising from the ground meristem, the cortex in many roots consists primarily of parenchyma cells with typical isodiametric dimensions and numerous intercellular spaces.

Keeping in mind the origin of the cortex, what other tissues besides parenchyma could you logically expect to find in the root cortex?

____

- - - - - - - - - - - - - - - - -

You could expect to find collenchyma and sclerenchyma since these, along with parenchyma, comprise the ground tissue which is produced by the ground meristem.

18. Cortical cells of the root often contain an abundance of food in the form of starch grains. Such starch-containing cells should remind you that one

of the functions of the root is ______________________.

What are three additional functions of roots? ______________________

____

- - - - - - - - - - - - - - - - -

food storage; absorption of water and minerals from the soil, conduction of water, and anchorage (any order)

19. The innermost limit of the cortex is marked by a ring of endodermis cells which differ from their neighboring parenchyma cells of the cortex in shape and in the manner in which their primary walls are formed.

A young endodermal cell is banded by a narrow thickened strip impregnated with waxy suberin which is an integral part of the primary wall. This constitutes the Casparian strip, which crosses the two transverse (top and bottom) walls and the two radial (side) walls, as illustrated below in a three-dimensional view of an endodermal cell.

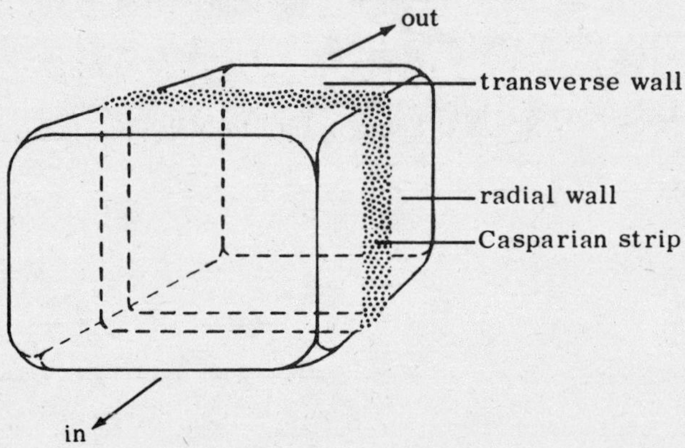

The innermost layer of the cortex is the ______________________,
whose cells possess thickened bands called ______________________.

- - - - - - - - - - - - - - - - -

endodermis; Casparian strips

20. Although the true function of the endodermis is not clear, the suberized Casparian strips appear to act as watertight seals around each cell of the endodermis. Therefore, any minerals which are dissolved in the soil water cannot move into the stele of the root without first passing through the protoplasm of the endodermal cells.

Due to the nature of protoplasm some minerals are barred entrance into the stele while others pass through readily. Thus, the endodermis may serve as a regulatory tissue regarding the passage of minerals.

With the passage of time, the walls of most endodermal cells become greatly thickened as an uneven deposition of suberin and cellulose forms secondary walls.

Name another plant tissue besides endodermis which is characterized by a large deposition of suberin in its walls. ______________________

- - - - - - - - - - - - - - - - -

cork tissue

21. We have concluded our study of the cortex with the description of the endodermis. Let us now consider the nature of the stele. Derived from the Greek word for a column, the stele may be visualized as a central column of tissues which are differentiated from the procambial portion of the apical meristem. The stele is composed of both vascular and nonvascular tissues.

Vascular tissue is an inclusive term for the two complex tissues, __________________________ and __________________________.

- - - - - - - - - - - - - - - - -

xylem, phloem (either order)

22. Until now, our discussion of the root has not distinguished between the monocots and dicots, for the anatomy of the epidermis and cortex of these two groups is similar. Anatomical differences between the two kinds of roots occur primarily in the stelar region. For that reason, we must consider the steles of dicots and monocots separately.

Why are the two evolutionary groups of plants called monocots and dicots? ____

____

- - - - - - - - - - - - - - - - -

The embryos of monocots have one cotyledon, while dicot embryos have two.

23. The stele of a typical dicot root is encircled by a single layer of parenchyma-like cells which constitutes the <u>pericycle</u>. Functionally, the pericycle is one of the most important tissues in the dicot root.

If the pericycle is the outermost layer of the stele, what layer of cells lies immediately external to it? ______________________________

- - - - - - - - - - - - - - - - -

The endodermis, which is the innermost layer of the cortex.

24. The center of the stele of the dicot root is usually occupied by a mass of xylem with radiating arms extending to the pericycle (see photograph). The phloem lies in somewhat circular patches alternating with the xylem arms. This alternate arrangement of xylem and phloem is typical of all roots.

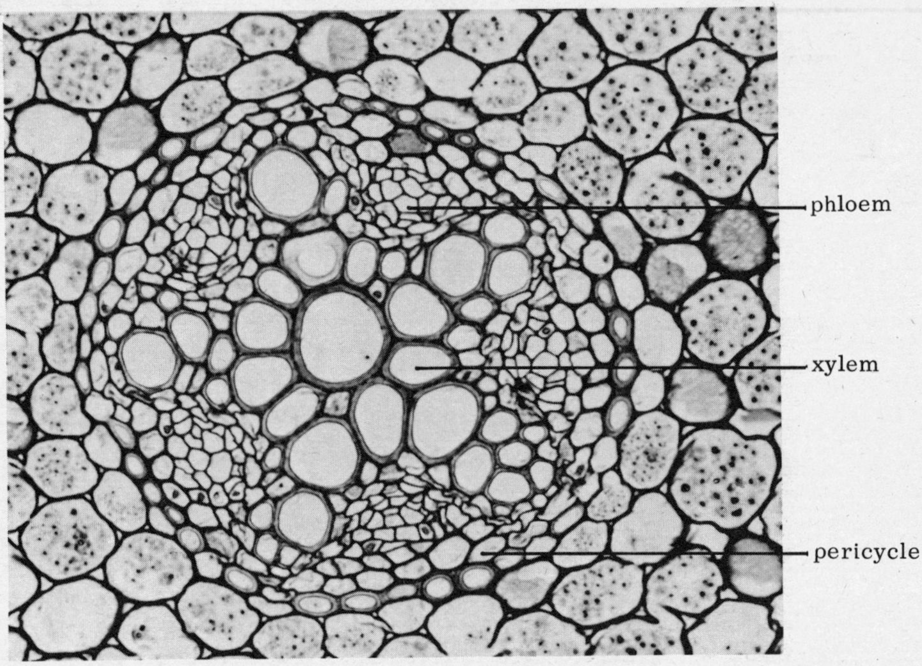

cross section of a stele of a buttercup root
(Courtesy Carolina Biological Supply Company.)

In the above photograph of the stele:

a. Which of the tissues of the stele are in direct contact with the pericycle? ____________

b. What substance moves through the centermost tissue? ____________

\- - - - - - - - - - - - - - - - - -

a. xylem and phloem; b. water (and minerals dissolved in it)

25. The number of xylem arms in the root may vary from two to many, depending on the species. On the next page, the xylem is labeled in the diagram of a dicot stele. Draw the location of the phloem.

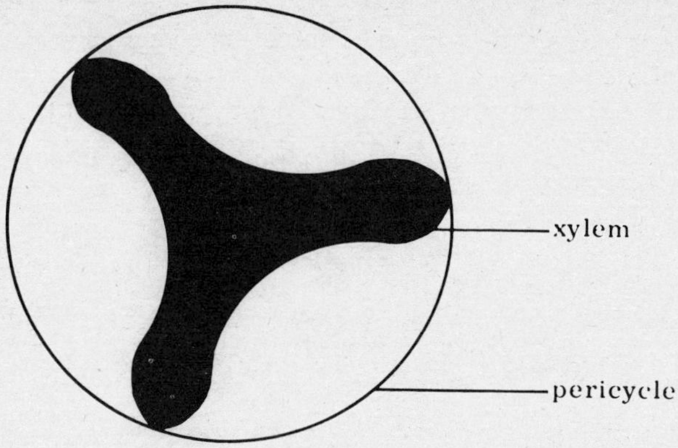

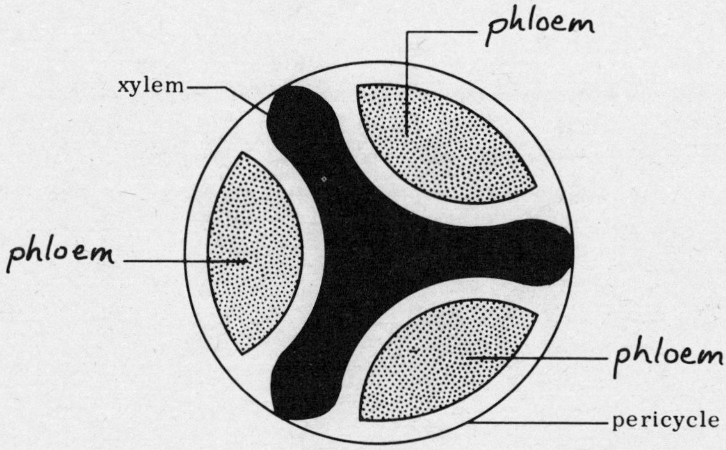

26. The xylem and phloem tissues of the root are separated by a single layer of procambial cells which may remain meristematic and form a cambium. If the root does not increase in diameter by cambial activity, the procambial cells form sclerenchyma or parenchyma. The following diagram shows the relationship of the potential cambial cells to the xylem and phloem.

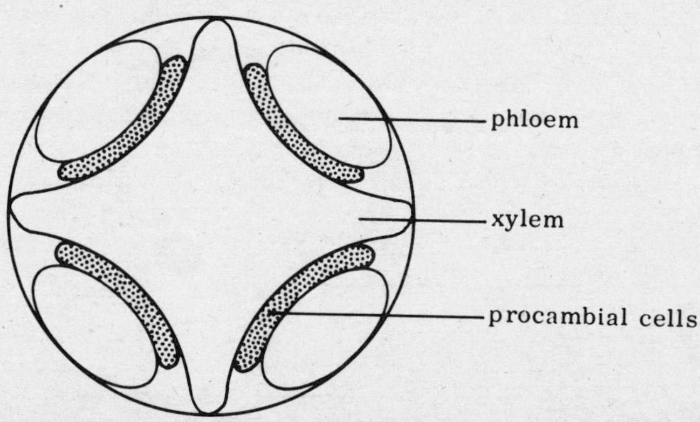

Keeping in mind the nature of the stele, place a check before the statements below which are correct.

________ a. If the meristematic region separating xylem and phloem develops into a cambium, the xylem and phloem are separated by a lateral meristem.

________ b. Xylem tissue forms a solid core in the stele while phloem tissue occurs as isolated strands.

________ c. Xylem, phloem, and the pericycle are formed from the procambium.

- - - - - - - - - - - - - - - - -

All three are correct statements.

27. You have learned something of the composition and arrangement of the tissues in the stele of a typical dicot root. Now let us consider the similarities and differences encountered in a monocot stele. In many, but not all, monocots the pericycle may be several layers thick. The cells often develop thick secondary walls and become sclerenchyma-like in older roots. A young pericycle occasionally contains isolated cells of xylem, phloem, or other specialized cell types and thus loses its homogeneous nature.

Can you suggest how a pericycle may be recognized as an anatomical entity in instances where the pericycle is several layers thick, or composed of several kinds of cells? ________________________________

____

- - - - - - - - - - - - - - - - -

The pericycle is recognized primarily by its relative position in the stele as well as by its origin from the procambium.

28. The center of a monocot stele may be varied, but the typical monocot stele is exemplified by corn in which the center is occupied by parenchyma cells. The term <u>pith</u> is used to describe the occurrence of ground tissue in the center of the stele. Therefore, the stele of a corn root as well as the steles of many other monocot roots possesses a pith derived from the procambium.

The kind of ground tissue comprising the root of corn is

________________________ tissue.

- - - - - - - - - - - - - - - - -

parenchyma

29. Ground tissue is a term used to describe those tissues derived from what portion of an apical meristem? ______________________

Ground tissue is composed of what three simple tissues? ____________ ______________________________ The parenchyma produced in the pith of a corn root differs from typical parenchyma in that pith parenchyma arises from what portion of the apical meristem?

____

- - - - - - - - - - - - - - - - - -

ground meristem; parenchyma, collenchyma, and sclerenchyma (any order); procambium

30. Study the photograph of a monocot root shown below.

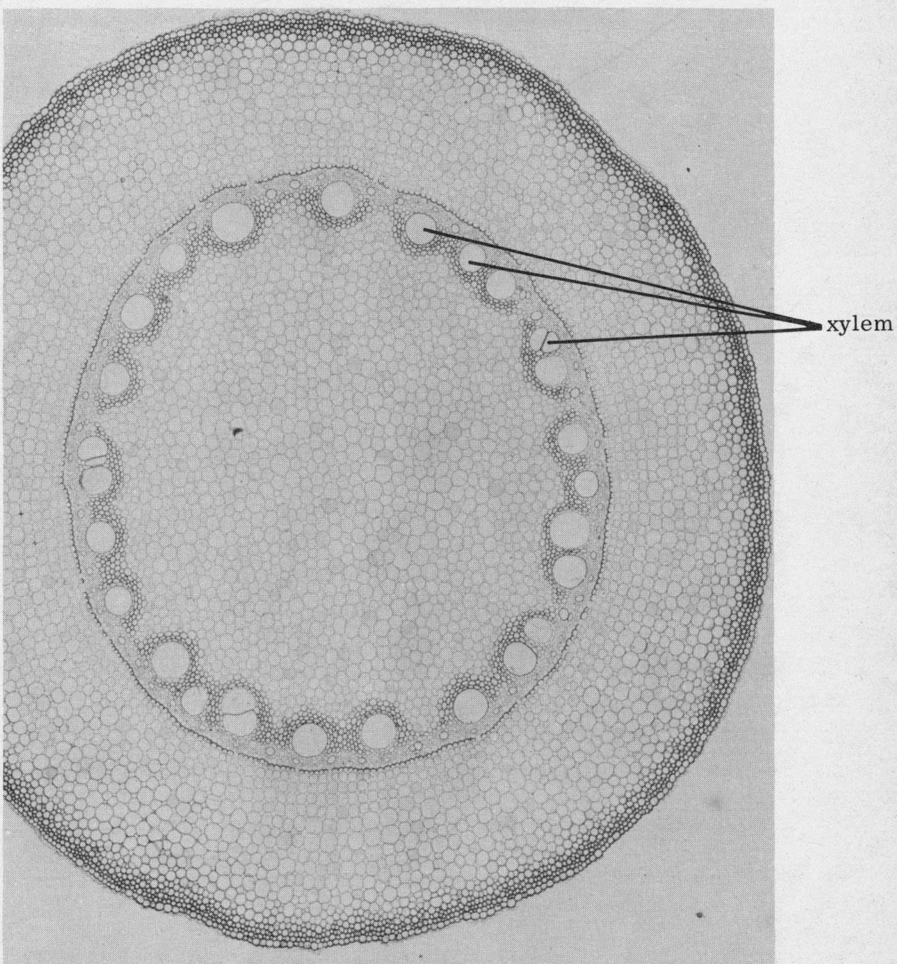

(Courtesy Carolina Biological Supply Company.)

Keeping in mind the relationship of xylem and phloem in a root, where would you expect to find phloem tissue in the monocot stele?

____

- - - - - - - - - - - - - - - - -

As in all roots, you would find the phloem alternating with the xylem units.

31. The diffuse nature of the vascular tissue in monocot roots, together with the tendency of nearby parenchyma cells to develop thickened secondary walls, make the identification of the vascular components more difficult than in dicot roots. The absence of a procambial layer of potential cambium between the xylem and phloem of monocots marks another great difference between monocot and dicot roots.

Label the parts of a monocot root in the following diagram.

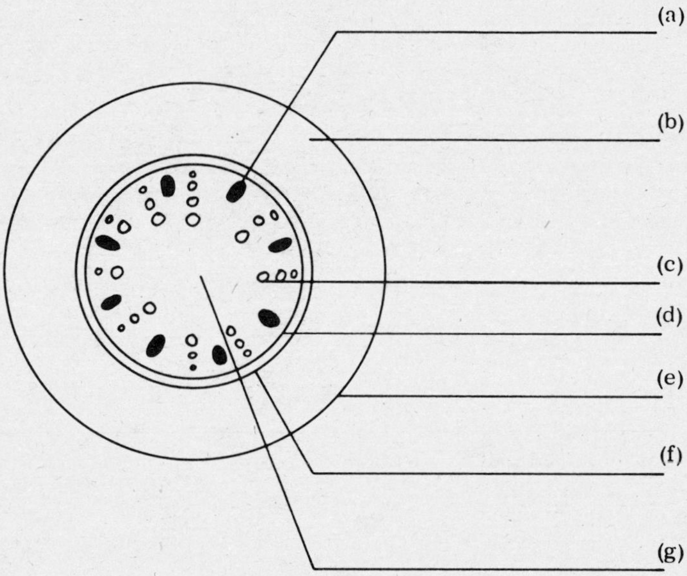

- - - - - - - - - - - - - - - - -

a. phloem; b. cortex; c. xylem (vessels); d. pericycle; e. epidermis; f. endodermis; g. pith

32. You learned earlier that roots may arise from three sources. These are:

____

____

- - - - - - - - - - - - - - - - -

radicle of embryo, other roots, and adventitiously (any order)

33. We can expect to find branch roots arising from the taproot. Let's see how this comes about. Branch roots, or secondary roots, arise in the same manner in both dicot and monocot roots. If we examine a cross section of a root once more, we may find areas in which a section of the pericycle has become meristematic. A tongue of newly formed cells pushes into the cortex and assumes the internal structure of a root. Thus, branch roots arise from the pericycle and grow through the cortex and epidermis before reaching the soil. The vascular elements of the newly formed root become continuous with those of the older root so that conduction of water and food may proceed uninterrupted in the young root.

The origin of branch roots is said to be endogenous (endo = within + gen = generating or forming). See illustration below. Can you explain why this term is appropriate? ______________________________

____

- - - - - - - - - - - - - - - - -

Branch roots form from the pericycle of the stele, thus arising deep within the root.

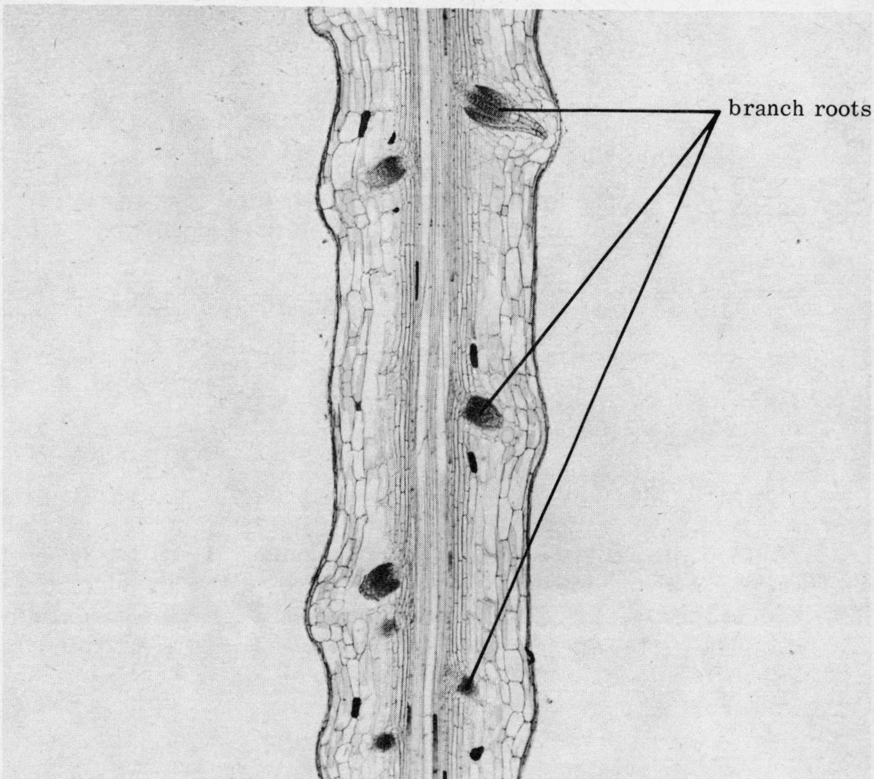

longitudinal section of a root showing origin of branch roots from the stelar region (Courtesy Carolina Biological Supply Company.)

34. You should not conclude that all dicot roots are structured in the manner previously described for the typical dicot. Some roots possess a pith and a few have multi-layered pericycles. Many variations occur in the steles of monocots. Occasionally, one or more xylem units differentiate in the center of the stele, while the appearance and number of radiating xylem arms vary greatly. You should concentrate, then, on the characteristics which unify all roots.

To summarize the generalizations made about monocot and dicot roots, match one of the following numbered choices with each pair of statements about roots.

1. Both statements I and II are true.
2. Statement I is true but II is false.
3. Statement I is false but II is true.
4. Both statements I and II are false.

________ a. (I) Characteristically dicot roots have a pericycle consisting of a single layer of cells; (II) the pericycle is the innermost region of the cortex.

________ b. (I) The anatomy of the epidermis and cortex of monocot and dicot roots are quite different, while (II) the stelar regions show no appreciable differences.

________ c. (I) The center of a cross section of a typical dicot root is made up of xylem, while (II) the center of a cross section of a typical monocot root is occupied by pith.

________ d. Branch roots: (I) originate in the same way in both monocots and dicots, (II) develop from the endodermis.

________ e. (I) In a dicot root phloem occurs in a solid central core and (II) the stele consists of the pericycle, xylem, and phloem.

- - - - - - - - - - - - - - - - -

a. 2; b. 4; c. 1; d. 2; e. 3

35. Lettered in the following list are some parts of a typical dicot root: (A) xylem, (B) cortex, (C) epidermis, (D) pericycle, (E) endodermis.

Suppose that you were the size of a bacterial cell and that you were entering a dicot root through a root hair. Trace your path from the exterior to the center of the root by listing the letters corresponding to the root parts in the sequence in which you would pass through them.

____

- - - - - - - - - - - - - - - - -

C—B—E—D—A

36. Finally, to summarize what you have learned about roots and the various components of roots, complete the following matching exercise.

________ a. Primary cell walls contain suberized thickenings called Casparian strips.		1. cambium
		2. cortex
		3. endodermis
________ b. Composed of pericycle, xylem, and phloem.		4. epidermis
		5. ground meristem
		6. pericycle
________ c. Meristem in dicot roots that develops between xylem and phloem and is responsible for increase in root diameter.		7. phloem
		8. pith
		9. stele
		10. xylem
________ d. Capable of giving rise to parenchyma, collenchyma, or sclerenchyma in root cortex.		
________ e. Vascular tissue which transports water and dissolved minerals to aerial plant parts.		
________ f. Cells from which branch roots arise.		
________ g. Parenchyma cells of a monocot root that are derived from procambium.		

- - - - - - - - - - - - - - - - -

a. 3; b. 9; c. 1; d. 5; e. 10; f. 6; g. 8

SELF-TEST

Before you proceed to Chapter 5, take this Self-Test to aid you in reinforcing what you have learned about roots in this chapter. All questions are to be answered in the light of information in Chapter 4 of this book. Answers are given at the end of this quiz.

1. List three or four general functions of roots.

   a. ______________________________

   b. ______________________________

   c. ______________________________

   d. ______________________________

2. List three ways in which roots may originate.

   a. ______________________________

   b. ______________________________

   c. ______________________________

3. Plants usually develop one of two kinds of root systems—a fibrous root system or a taproot system. Indicate the type of root system described in each case below.

   ____________ a. Root system characteristic of most monocots.

   ____________ b. Root system characteristic of a dandelion.

   ____________ c. Root system in which a primary root stores an abundant supply of food.

   ____________ d. Root system especially capable of holding the soil on a hillside.

   ____________ e. Root system capable of reaching water supplies deep in the soil.

   ____________ f. Root system in which primary, secondary, and tertiary roots are of approximately equal diameter.

4. Match the numbered root regions with the appropriate description.

   ______ a. Root hairs externally distinguish this part of the root.

   ______ b. Region where cell number is rapidly increasing.

   ______ c. Region where cell size is rapidly increasing.

   1. rootcap
   2. meristematic region
   3. region of elongation
   4. region of differentiation

(continued)

_______ d. Cells that protect the growing root tip.

_______ e. Region where epidermis, cortex, and stele first become evident.

5. Match the various root components with their descriptions.

_______ a. Cells from which branch roots arise.

_______ b. Vascular tissue which transports water and dissolved minerals to aerial plant parts.

_______ c. Capable of giving rise to parenchyma, collenchyma, or sclerenchyma in root cortex.

_______ d. Parenchyma cells of a monocot root that are derived from procambium.

_______ e. Meristem in dicot roots that develops between xylem and phloem and is responsible for increase in root diameter.

_______ f. Composed of pericycle, xylem, and phloem.

_______ g. Primary cell walls contain suberized thickenings called Casparian strips.

1. stele
2. endodermis
3. epidermis
4. xylem
5. phloem
6. pith
7. cambium
8. ground meristem
9. pericycle
10. cortex

Answers

Compare your answers to the Self-Test with the correct answers given below. If all of your answers are correct, you are ready to go on to the next chapter. If you missed any questions, review the frames indicated in parentheses following the answer before you go on.

1. a. anchorage; b. absorption of water and minerals; c. conduction of water and dissolved minerals; d. food storage (any order) (frame 1)

2. a. from the radicle of the embryo; b. from other roots; c. adventitiously from stems, etc. (any order) (frame 2)

3. a. fibrous (frames 5 and 6)
   b. taproot (frame 7)
   c. taproot (frame 4)
   d. fibrous (frame 6)
   e. taproot (frame 6)
   f. fibrous (frame 3)

4. a. 4 (frame 13)
   b. 2 (frame 9)
   c. 3 (frame 11)
   d. 1 (frame 9)
   e. 4 (frame 14)

5. a. 9 (frame 33)
   b. 4 (frame 24)
   c. 8 (frames 10 and 17)
   d. 6 (frame 28)
   e. 7 (frame 26)
   f. 1 (frames 23 and 24)
   g. 2 (frames 19 and 20)

CHAPTER FIVE
The Stem

1. We have traced the development of the root from its origin in the radicle of the seed embryo to its maturity. Let us now consider the formation and development of the stem as it arises from the apical meristem or epicotyl of the embryo.

   Because the tissues of the root (which we reviewed in Chapter 4) have developed from an apical meristem, they are considered to be what kind of tissues? ______________________

   - - - - - - - - - - - - - - - - -

   primary

2. The stem of a plant performs several important functions, such as: (1) providing support for the leaves and flowers, (2) functioning as a food manufacturing organ, (3) conducting food and water between the leaves and roots, and (4) acting as a food storage organ.

   What functions of the stem are duplicated by the root?

   ____

   ____

   - - - - - - - - - - - - - - - - -

   The stem and the root both store food and both act as organs of conduction.

3. The stem differs from the root both externally and internally, although both of them arise from apical meristems and together comprise the axis of the plant. The external appearance of a stem is best understood by examining a woody twig from a tree or shrub. The tip or youngest part of the stem bears the <u>terminal bud</u> which is made up of the apical meristem and surrounding leaves. Some of the oldest leaves become modified into <u>bud scales</u> which protect the bud during the dormant season.

From a functional standpoint these bud scales correspond to what part of a root tip? ________________________

- - - - - - - - - - - - - - - - -

the rootcap, which functions as a protective structure around the tip of the root in the same manner that the bud scales protect the tender tissues in the stem tip

4. Other buds occur along the sides of the stem and represent lateral buds. Growth of the lateral buds results in branching of the main stem. Because these branches occur in a superficial position, the stem is said to branch exogenously (exo = outside + gen = generating or forming). Here, again, is another difference between roots and stems, for branching of the root occurs ________________, while branches of the stem form ________________.

- - - - - - - - - - - - - - - - -

endogenously; exogenously

5. The leaves arise from the stem immediately below the points of lateral bud attachment. When the leaves are shed in the fall, their former positions on the stem are marked by leaf scars.

Using the information presented in these last three frames, label the following drawing of a twig.

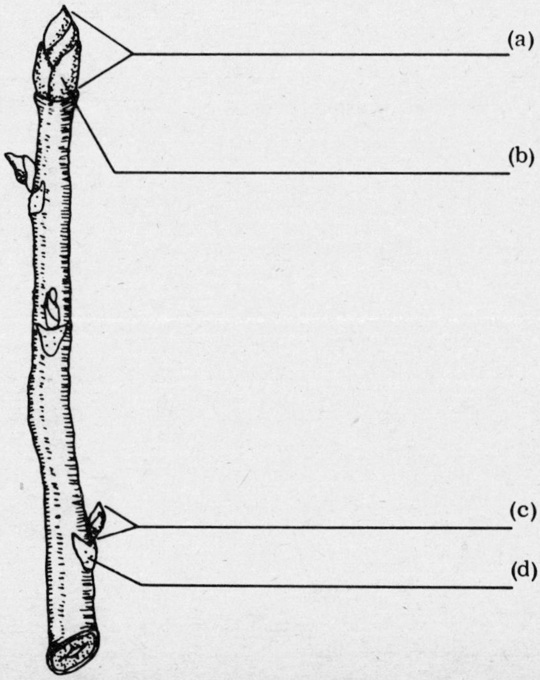

- - - - - - - - - - - - - - - - - -

a. terminal bud; b. bud scale; c. lateral bud; d. leaf scar

6. The regions of the stem to which the buds and leaves are attached are called nodes. Since leaves and buds occur on the stem at definite intervals, the stem is thus composed of a series of nodes. The portion of the stem which lies between two successive nodes comprises the internode.

Label nodes and internodes on the stem diagram below.

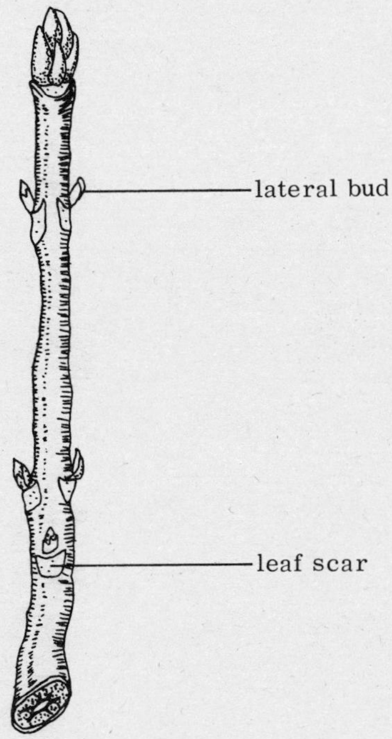

- - - - - - - - - - - - - - - - - -

Look on the next page for the diagram with labels as you should have written them.

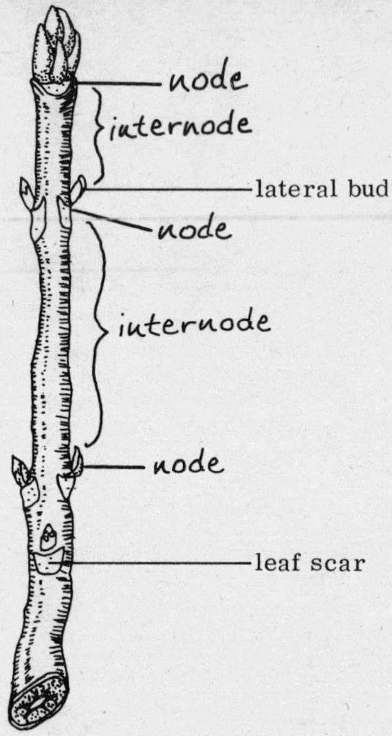

7. You have learned several new terms pertaining to the external nature of the stem. To review these terms, match the following words with their corresponding definitions.

________ a. Mark left on the stem surface where a leaf was attached.		1. bud scale
________ b. Bud which increases the length of the main stem.		2. terminal bud
________ c. Modified leaf which protects an apical meristem.		3. exogenous
________ d. Portion of the stem to which leaves and buds are attached.		4. node
________ e. Bud which produces branches of the stem.		5. leaf scar
________ f. Portion of the stem between two points of leaf attachment.		6. lateral bud
________ g. Term used to describe a superficial branching of the plant stem.		7. internode

- - - - - - - - - - - - - - - - -

a. 5; b. 2; c. 1; d. 4; e. 6; f. 7; g. 3

8. Let us assume that you have made a longitudinal section of the terminal bud. A microscopic examination of the bud reveals the manner in which the stem arises from the apical meristem. The following photograph illustrates the appearance of a typical bud viewed in longitudinal section.

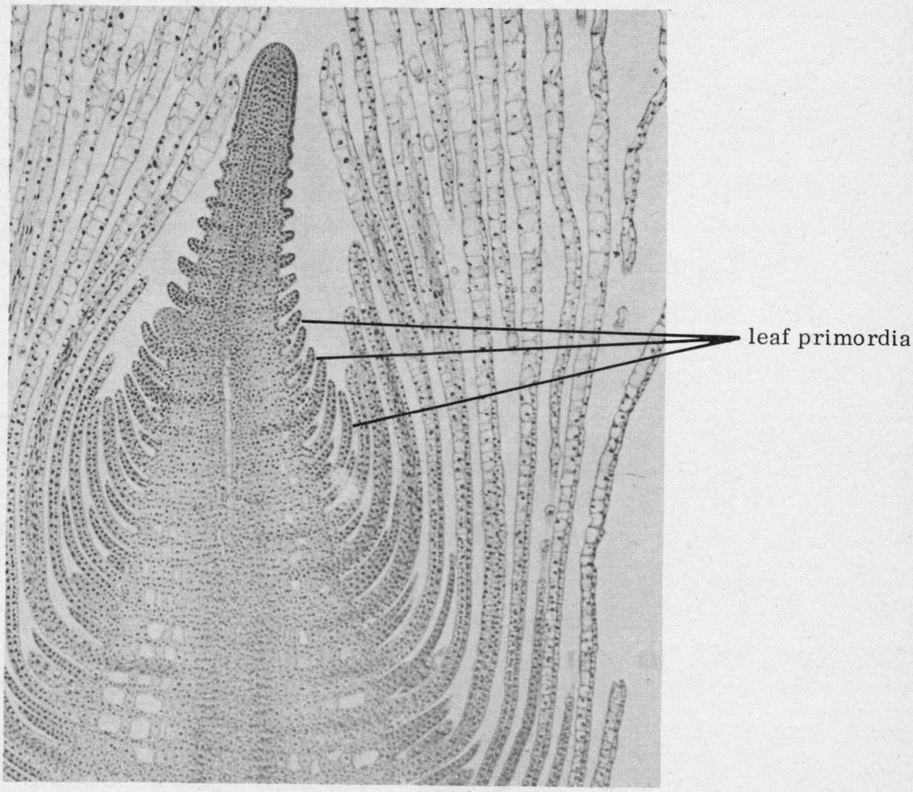

(Courtesy Carolina Biological Supply Company.)

Differentiation of tissues begins early from the extreme tip of the bud. From the apical meristem itself proliferation of cells establishes leaf primordia (singular, primordium) which project as lateral outgrowths from the stem. The cells comprising the young stem differentiate into protoderm, ground meristem, and procambium.

As in the root tip, the protoderm eventually gives rise to the ______________________________ of the mature stem. Ground tissue is produced by the ____________________________, and the procambium differentiates into ____________________________ tissue.

- - - - - - - - - - - - - - - - -

epidermis; ground meristem; vascular tissue (xylem and phloem)

9. Subsequent cell division and enlargement of leaf primordia cells result in the growth of the young leaves. Which of the leaf primordia illustrated in frame 8 are older, the ones nearest the tip of the bud or those nearest its base? ______________________

- - - - - - - - - - - - - - - - -

those nearest the base of the bud

10. While the leaf primordia are attached to the nodal areas of the stem in theory, in practice it is difficult to identify them because of the absence of well defined internodes. Cell division and elongation occur in all regions of the bud, but prolonged activity in the internodes long after the bud ceases to exist results in a profound separation of the nodes of the stem.

A botanist who is attempting to identify an unknown tree frequently needs to know the exact number of leaves which are attached to each node of the stem. Would the botanist stand a better chance of obtaining that information if the plant were examined in early spring as the buds open, or if it were examined later in the season? Justify your answer.

____

____

- - - - - - - - - - - - - - - - -

later in the season after the internodes have had time to elongate, making it possible to accurately count the number of leaves at each node

11. We have examined the prominent external features of a woody stem, including the nodes, internodes, terminal buds, and lateral buds. Many stems die at the end of the first growing season, however, and they remain soft-stemmed or herbaceous with their apical meristems lacking protective bud scales.

Suppose you were given a plant structure and asked to decide whether it was a stem or root based on external characteristics. What features would you consider in making your decision? ______________________

____

- - - - - - - - - - - - - - - - -

The presence of nodes, internodes, leaves or leaf scars, and buds at nodes would indicate that the structure was a stem.

12. Let us now consider the internal structure of an herbaceous stem. Since the structure of a dicot stem differs from that of a monocot, we must consider the two separately. An examination of a cross section of a young

dicot stem at the internode a short distance away from the apical meristem reveals that the outermost layer of cells has differentiated into an epidermis. As in the root, the epidermis serves as a protective layer.

What primary meristematic region produced the epidermis?

____

- - - - - - - - - - - - - - - - -

The protoderm has differentiated into the epidermis.

13. The epidermis of the stem differs in several ways from that of the root. The stem epidermis is characterized by extensive cutinization of the cell walls, a cuticle of varying thickness, numerous stomata, and the presence of various kinds of hairs which are often secretory in nature.

    Stinging nettle causes painful irritation when it is merely brushed against the skin. In the light of what has been discussed in this frame, can you guess how stinging nettle transmits its irritant to the skin?

    ____

    - - - - - - - - - - - - - - - - -

    Short, stiff hairs on the epidermis of stinging nettle act as tiny hypodermic needles to secrete their irritant into the skin.

14. A certain mutant gene causes the stems of tomato plants having this gene to be "wooly"—heavily covered with hairs. What cellular component of the stem would you expect to produce these hair cells? ____________

    ____

    - - - - - - - - - - - - - - - - -

    They are part of the epidermis.

15. A stem cortex, which is generally narrower than that of a root, is differentiated from the ground meristem and lies immediately below the epidermis of the stem.

    Since it is formed from ground meristem, you should expect the cortex to be composed of some or all of what three kinds of tissues?

    ____

    - - - - - - - - - - - - - - - - -

    parenchyma, collenchyma, and sclerenchyma (any order)

16. The stem cortex presents a more heterogeneous picture than that of the root because of the variety of cells which comprise it. Parenchyma is

the most common tissue in the cortex, and in herbaceous green stems, chlorophyll forms in the superficial parenchyma cells. The term chlorenchyma is sometimes applied to these cells. In acting as food storage units and as potentially meristematic cells, parenchyma forms an important constituent of the cortex.

What tissue primarily constitutes the root cortex? ______________________________

- - - - - - - - - - - - - - - - - -

parenchyma

17. When collenchyma is present, it usually comprises the outermost region of the cortex and forms several layers of cells just below the epidermis. Additional support of the stem may be provided by sclerenchyma, usually in the form of fibers, but sclereids occasionally occur.

Label the tissues in the photograph below of the cortex of the stem of a squash plant.

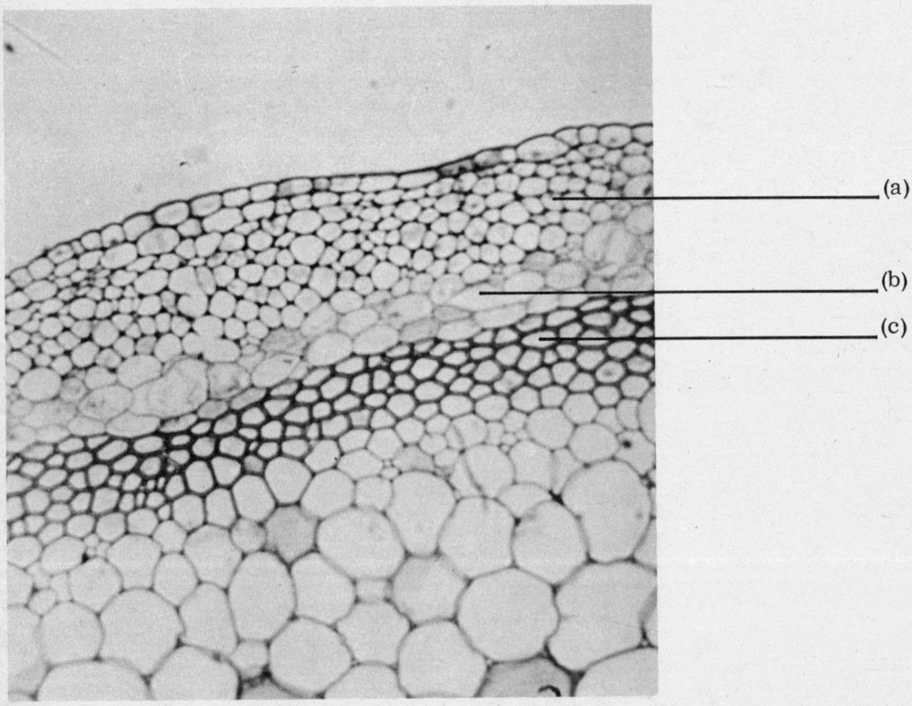

- - - - - - - - - - - - - - - - - -

a. collenchyma; b. parenchyma; c. sclerenchyma (fibers)

18. You learned in the previous chapter that the innermost layer of the root cortex was marked by the occurrence of a layer of cells called the

________________________.

- - - - - - - - - - - - - - - - -

endodermis

19. With very few exceptions, the endodermis is lacking in stems. The pericycle which serves as an important tissue in roots is also lacking in stems. Since stems lack a pericycle and an endodermis, can you distinguish between the cortex and the stele? ____________________

____

- - - - - - - - - - - - - - - - -

No. The absence of both endodermis and pericycle makes the establishment of the inner and outer boundaries of the cortex and stele impossible.

20. The central region of the dicot stem is usually occupied by a pith composed of parenchyma differentiated from the ground meristem. In contrast, the parenchyma tissue which occasionally forms in the well-defined stele of the root is derived from the procambium as is all of the stelar tissue. Because of the lack of anatomical boundaries between the cortex and the central portion of the stem, together with the nature of the origin of the pith, the stelar concept will be omitted in this discussion of stem anatomy.

a. What tissue usually occupies the center of a dicot root? __________

____

b. What tissue usually occupies the center of a dicot stem? __________

____

- - - - - - - - - - - - - - - - -

a. xylem tissue; b. pith parenchyma

21. The procambium of a very young stem commonly appears as isolated strands running vertically through the ground meristem. Thus, the differentiation of vascular tissue from the procambium results in the establishment of bundles of xylem and phloem arranged in a ring around a central column of parenchyma (see diagram on next page).

The strands of vascular tissue, or vascular bundles, are separated by radiating arms of the pith. These radiating arms of parenchyma cells constitute the pith rays which serve as regions of storage and lateral conduction of materials between the pith and the cortex.

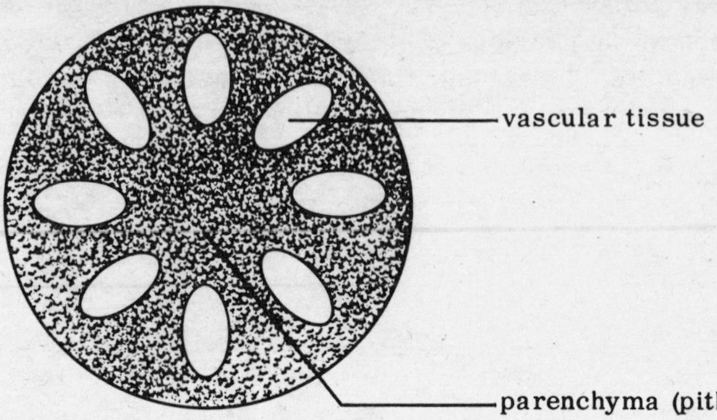

Label vascular bundles, cortex, epidermis, pith, and pith rays in the following photograph.

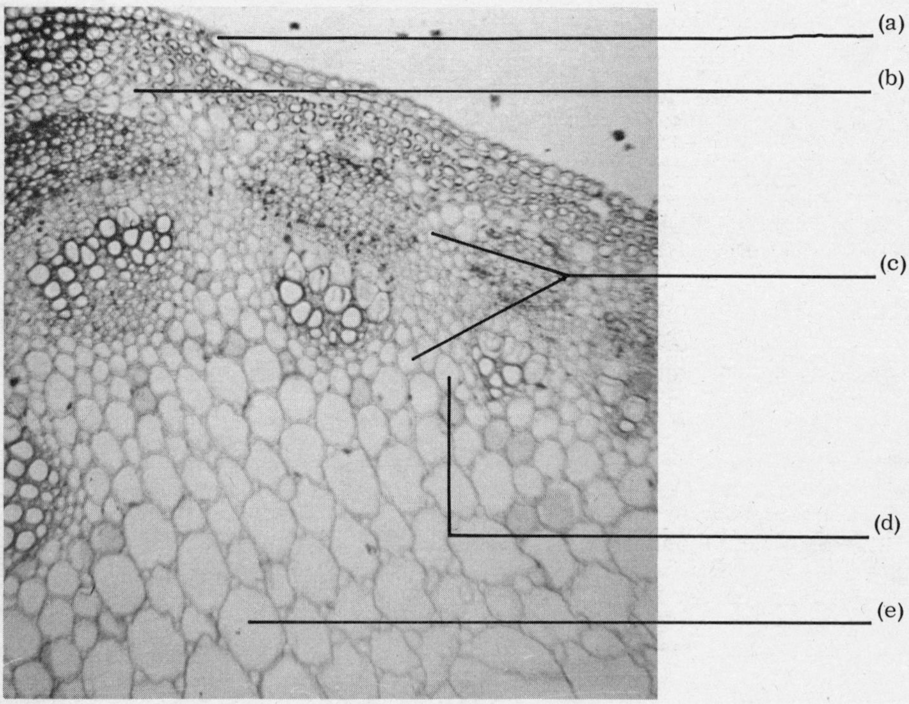

---

a. epidermis; b. cortex; c. vascular bundle; d. pith ray; e. pith

22. A typical dicot vascular bundle is composed of a region of xylem lying toward the inside of the stem with phloem tissue lying toward the outside. Such an arrangement of xylem and phloem constitutes a collateral bundle, in contrast to the less common bicollateral bundle in which the xylem lies between inner and outer phloem strands.

Label xylem and phloem in the two diagrams which follow. The first diagram-photograph pair represents collateral bundles, while the second diagram and photograph represent bicollateral bundles.

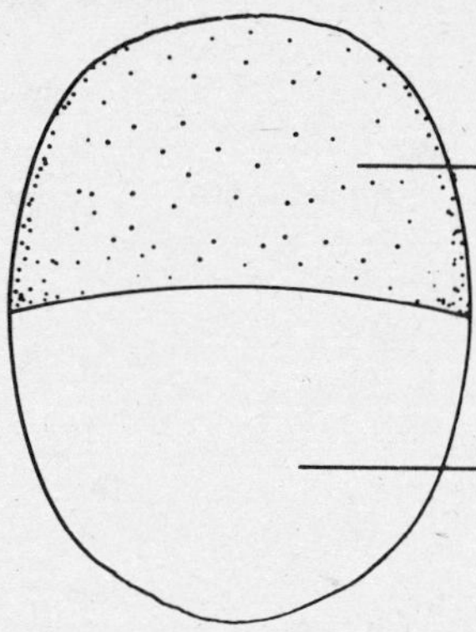

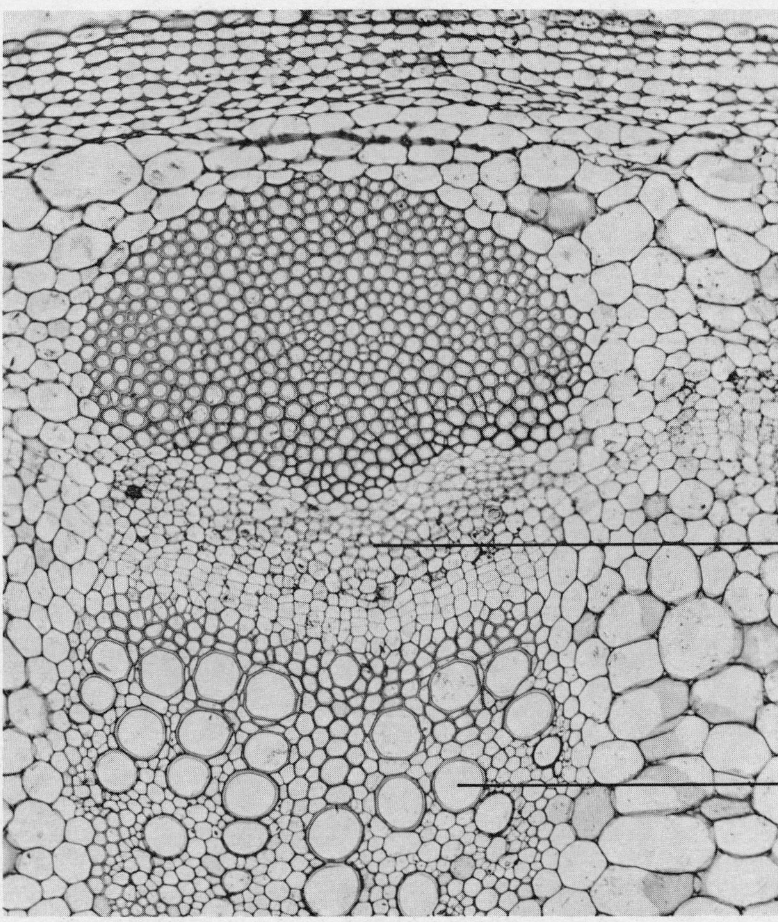

(Courtesy Carolina Biological Supply Company.)

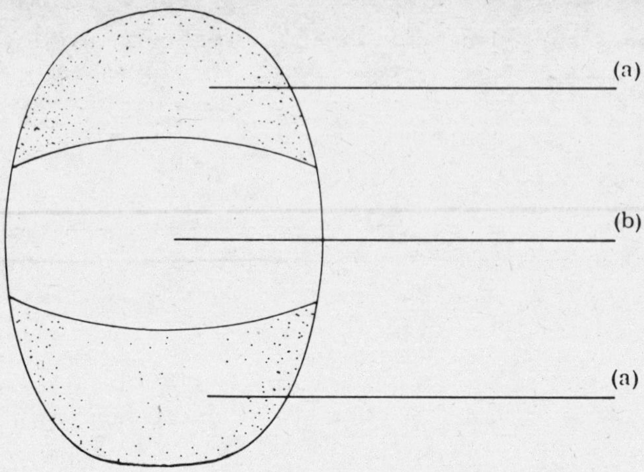

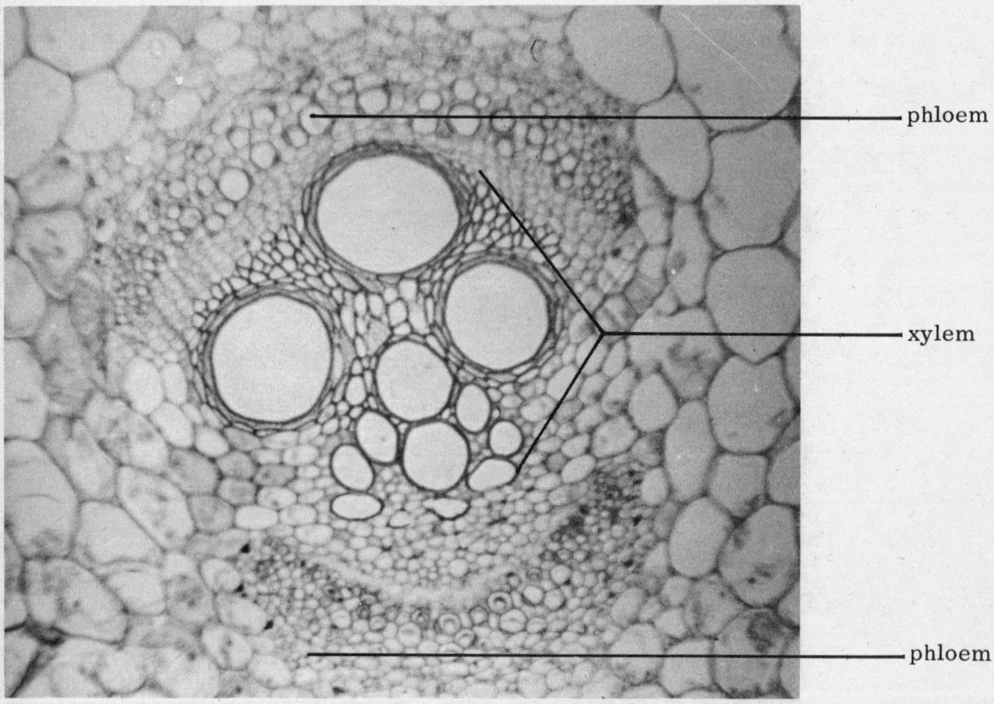

\- - - - - - - - - - - - - - - - - -

a. phloem; b. xylem

23. In vascular bundles of the collateral type, the xylem and phloem are separated by an undifferentiated layer of the procambium. In plants which live for a single season, the procambium's identity remains unchanged. In woody plants, however, the procambium develops into a lateral meristem called the vascular cambium.

In what region or regions would you expect to find procambium in a bicollateral bundle? ______________________

____

- - - - - - - - - - - - - - - - -

You may logically expect to find two procambiums separating the xylem from the external and internal patches of phloem. Usually, when two procambiums do appear the inner one is less obvious.

24. Perhaps you have noticed that the arrangement of xylem and phloem in a stem is markedly different from that of a root. When a radial line is drawn outward from the pith of a stem, that line passes through both xylem and phloem. Would a similar line drawn from the center of a root transect both xylem and phloem? Why? ______________________

____

- - - - - - - - - - - - - - - - -

A single radial line drawn across the root would not necessarily transect both tissues because xylem and phloem lie in an alternate pattern in the root.

25. Let us now turn our attention to an herbaceous monocot stem. Although monocots vary a great deal in stem arrangement, a common stem type is exemplified by corn. Examination of a cross section of a corn stem reveals that the vascular bundles are scattered at random, and there is no well-defined pith (see photograph at top of next page).

The corn stem is often described as "pithy." Is this a correct term from a botanical point of view? Why? ______________________

____

- - - - - - - - - - - - - - - - -

No. It is not an accurate term because a true pith does not exist due to the scattered nature of the vascular bundles.

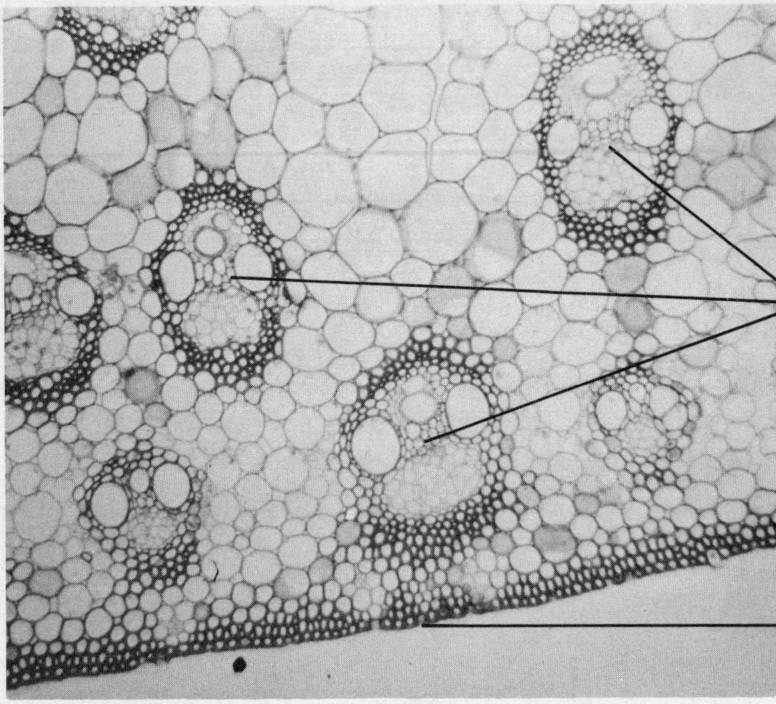

26. Each vascular bundle of the corn stem presents a characteristic appearance because of the arrangement of the largest xylem units. Close examination, however, reveals that the bundles are of the collateral type (see photograph below).

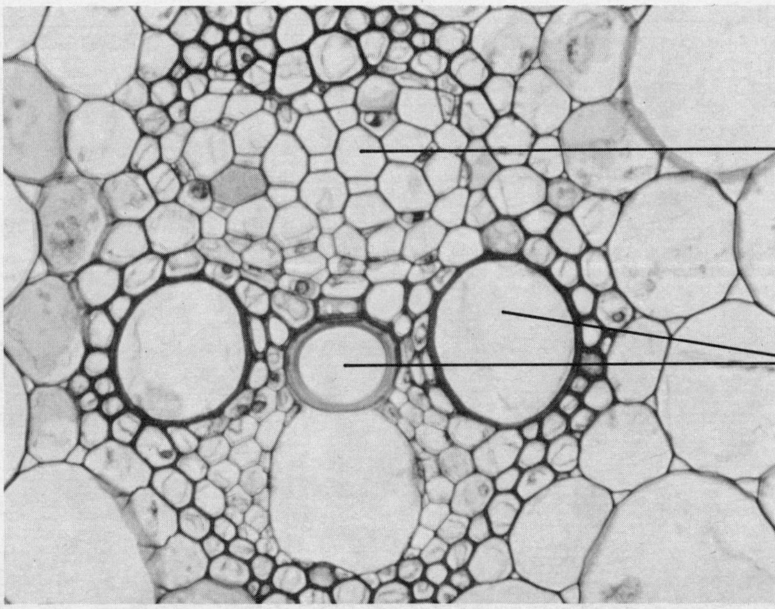

(Courtesy Carolina Biological Supply Company.)

Although the bundles themselves are scattered at random, the orientation of the xylem and phloem is such that the phloem lies external to the xylem. The collateral monocot bundle differs from the dicot bundle by the absence of procambium. With the exception of a few tropical forms, secondary growth does not occur in monocot stems due to the absence of a cambium. Therefore, monocot stems do not increase appreciably in diameter.

How does the absence of a vascular cambium in the monocot stem compare with the condition in the monocot root? ____________________

____

- - - - - - - - - - - - - - - - - -

A vascular cambium is generally absent in both roots and stems of monocots.

27. Throughout this chapter we have been comparing and contrasting roots and stems. To summarize the points made, complete the following exercise, indicating whether each statement is:

    1. true of stems only
    2. true of roots only
    3. true of both roots and stems

________ a. Possess nodes and internodes.

________ b. Protoderm gives rise to epidermis.

________ c. Possess a clearly defined pericycle.

________ d. Possess a cortex.

________ e. Possess vascular tissue.

________ f. Epidermis extensively cutinized and contains stomata.

________ g. Branch exogenously.

- - - - - - - - - - - - - - - - - -

a. 1; b. 3; c. 2; d. 3; e. 3; f. 1; g. 1

28. This chapter treats herbaceous dicot and monocot stems separately since there are some distinct differences between them. Indicate whether the following statements are true of monocot stems (M), dicot stems (D), or of both (D & M).

________ a. Vascular bundles form a ring around a centrally located pith.

________ b. Vascular bundles lack a procambium.

________ c. Stem has the potential to increase appreciably in diameter.

(continued)

________ d. Have both xylem and phloem.

________ e. Stem lacks a well-defined pith.

________ f. Stem covered by epidermis.

- - - - - - - - - - - - - - - - -

a. D; b. M; c. D; d. D & M; e. M; f. D & M

29. Most soft-stemmed herbaceous plants, such as petunias and marigolds, are classed as annuals because they are unable to live through the winter. Thus, annuals must produce their seeds during a single growing season before they succumb to the autumn frost. In contrast, perennials survive for many years and reproduce by seeds annually. The majority of perennials possess freeze-resistant woody stems which remain above ground, and from which the leaves are shed as cold weather approaches.

    A maple tree and a lilac bush are alike in that they can be classed as (annuals/perennials) ________________________. Are their stems herbaceous? ____

- - - - - - - - - - - - - - - - -

perennials; no, they are woody

30. A few perennials, especially monocots, remain herbaceous. Although their soft aerial stems and leaves die back to the ground every winter, their fleshy underground stems survive to send new aerial stems and leaves above ground the following year.

    Peony stems are killed by frost every year, yet new growth appears in the same area each spring. How many kinds of stems does the peony possess? ____

- - - - - - - - - - - - - - - - -

The peony has two kinds of stems. Annual stems which are herbaceous occur above ground, whereas fleshy perennial stems are found underground.

31. Let us consider a common perennial such as the spring crocus. In early March, clusters of bright flowers and grasslike leaves emerge from the cold earth. After a short period of blooming and seed production, the plant dies back to the ground. Below ground, however, the crocus stem persists as a vertical fleshy and solid structure. Such a modified underground stem is called a corm.

    New buds from the surface of the corm are responsible for the growth above ground the following spring (see photograph on next page).

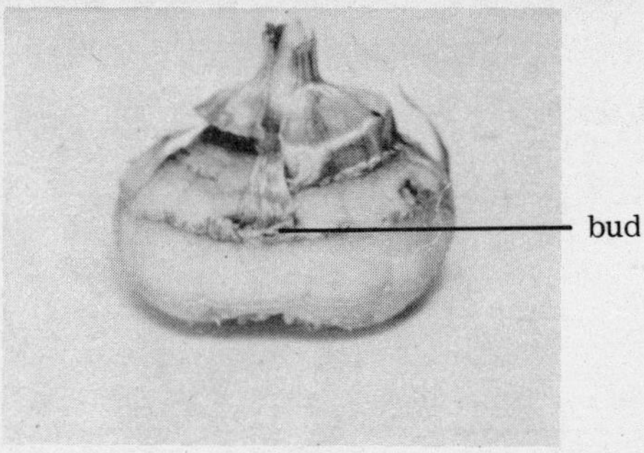

Which arrangement of the vascular tissue would you expect to find in a crocus corm?

________ a. Xylem and phloem arranged in bundles.

________ b. Xylem and phloem lying in an alternate pattern around the center of the corm.

- - - - - - - - - - - - - - - - - -

a, because the corm is a modified stem

32. A <u>bulb</u> is another fleshy underground structure which superficially resembles a corm. A longitudinal section of the bulb reveals a small stem whose nodes are separated by extremely short internodes. Fleshy leaves with an abundance of stored food are attached to each of the nodes. The terminal bud on the stem may send stems and leaves above ground when growth resumes (see diagram).

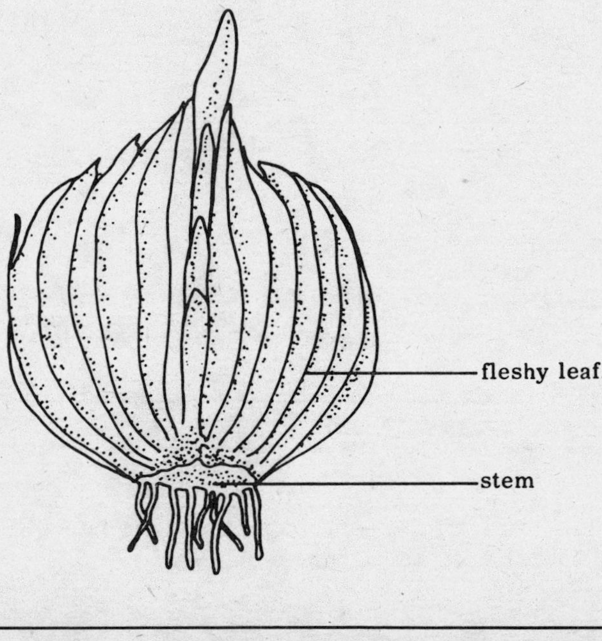

Which of the two underground structures more closely resembles a bud, a corm or a bulb? Why? ______________________________

____

- - - - - - - - - - - - - - - - -

The bulb closely resembles a bud with telescoped nodes from which are borne leaves. A corm lacks fleshy leaves but is covered with dry scale-like leaves.

33. Many herbaceous perennials produce a rhizome which may be described as a relatively unthickened horizontal underground stem. Lateral buds on the nodes give rise to new growth (see photograph).

iris rhizome

Many "weedy" grasses in a lawn form extensive rhizomes. Why is spot digging an ineffective method of ridding the lawn of unwanted weeds of this type? ______________________________

____

- - - - - - - - - - - - - - - - -

The rhizome has in all probability grown so far laterally that it is impossible to remove all of it by a single digging.

34. Greatly enlarged underground stems occasionally form at the ends of rhizomes in certain plants such as the Irish potato. These enlarged stems are known as tubers. Largely because of their abundant stored food and active buds, tubers produce typical aerial stems. The following photograph of the Irish potato tuber reveals that it possesses stem characteristics discussed in the early frames of this chapter.

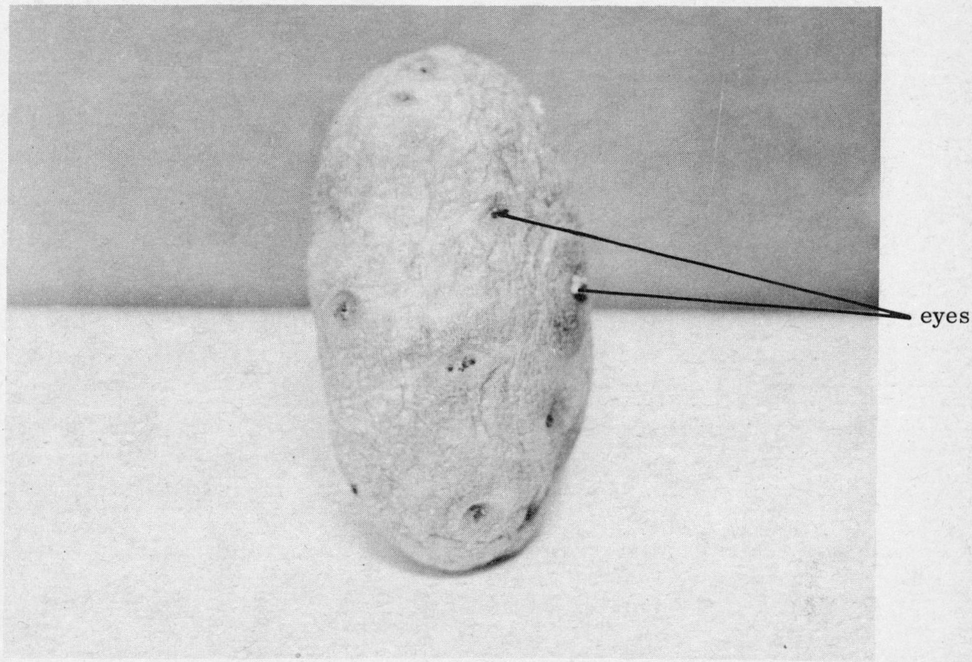

Irish potatoes are propagated by cutting tubers into pieces so that each portion contains one or two "eyes." When the piece is planted, a stem emerges from the eye. Can you guess what constitutes the "eye" of a potato tuber? ______________________________

- - - - - - - - - - - - - - - - - -

The eye is in reality a node from which one or more buds arise.

35. We have emphasized the fact that the Irish potato has more than one kind of stem. In fact, it has three different kinds of stems. Can you name these three kinds? ______________________________

____

- - - - - - - - - - - - - - - - - -

aerial stems, rhizomes, and tubers (any order)

36. To help you reinforce your understanding that corms, bulbs, rhizomes, and tubers are stems and not roots, list two or three characteristics that these structures possess that clearly identify them as stems. __________

____

____

Why do you think they are often confused with roots? ______________

____

- - - - - - - - - - - - - - - - - -

Corms, bulbs, tubers, and rhizomes possess such stem characteristics as nodes, internodes, buds, leaves, and leaf scars.
Corms, bulbs, tubers, and rhizomes are often mistaken for roots by non-botanists because they grow beneath the soil surface.

SELF-TEST

Before you proceed to Chapter 6, take this Self-Test to aid you in summarizing and reinforcing what you have learned about stems in this chapter. All questions are to be answered in the light of information in Chapters 4 and 5 of this book. Answers are given at the end of this quiz.

1. List four functions a stem may have. ______________________________

____

____

2. Roots and stems have been compared and contrasted throughout Chapter 5. Review some of the points made by completing the following exercise. Indicate whether each statement is:

   1. true of stems only
   2. true of roots only
   3. true of both roots and stems

   ________ a. Characteristically possess an endodermis.

   ________ b. Dicots possess a well-defined stele.

   ________ c. Dicots possess pith parenchyma cells.

   ________ d. Cortex contains chlorenchyma, collenchyma, and sclerenchyma.

   ________ e. Dicots possess pith rays.

   ________ f. Possess epidermis derived from protoderm.

   ________ g. Possess vascular bundles consisting of xylem and phloem.

3. Indicate whether the following statements are true of monocot stems (M), dicot stems (D), or both monocot and dicot stems (M & D).

   ________ a. Vascular bundles scattered throughout the stem.

   ________ b. Vascular bundles possess a vascular cambium.

   ________ c. Stem has a well-defined pith.

   ________ d. Stem has little potential to increase appreciably in diameter.

   ________ e. Stem covered by epidermis.

4. On the next page, label the lettered items in the diagram showing the external features of a woody stem.

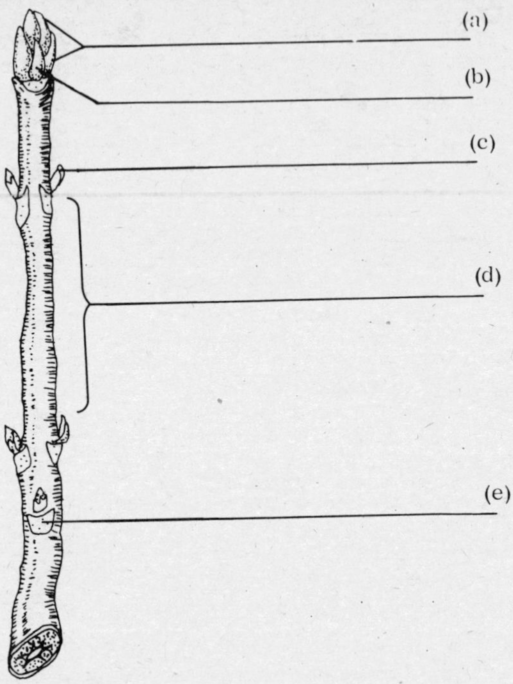

5. A number of common garden plants possess modified stems—corms, bulbs, tubers, or rhizomes. Identify the specific kind of modified stem to which each of the following descriptions refers.

____________________ a. The iris and canna are two garden plants, valued for their flowers, that produce horizontal underground stems with lateral buds at the nodes. These stems are probably examples of ____.

____________________ b. An onion is a plant with an underground stem which is flattened with short internodes and attached fleshy leaves surrounding a terminal bud. The onion thus is an example of a plant with a(n) ____.

____________________ c. The gladiolus is a garden plant having large, showy, colorful flowers. The stem of the gladiolus is a solid, fleshy underground structure that is called a(n) ____.

____________________ d. The Irish potato plant possesses enlarged, fleshy, underground stems that contain abundant stored food. These stems are called ____ and they possess active buds which may sprout and give rise to aerial stems.

6. Since corms, bulbs, rhizomes, and tubers grow beneath the soil surface, many nonbotanists think of them as roots. What features of these structures enable them to be recognized as stems? ______________________

____

____

Answers

Compare your answers to the Self-Test with the correct answers given below. If all of your answers are correct, you are ready to go on to the next chapter. If you missed any questions, review the frames indicated in parentheses following the answer before you go on.

1. conduction of food and water, support for leaves and flowers, storage of food, food manufacture (any order) (frame 2)

2. a. 2 (frame 18)
   b. 2 (frames 19 and 20)
   c. 1 (frame 20)
   d. 1 (frames 16 and 17)
   e. 1 (frame 21)
   f. 3 (frame 12)
   g. 1 (frame 21)

3. a. M (frame 25)
   b. D (frames 23 and 26)
   c. D (frames 20 and 25)
   d. M (frame 26)
   e. M & D (frames 12 and 28)

4. a. terminal bud; b. bud scale; c. lateral bud; d. internode; e. leaf scar (frames 5 and 6)

5. a. rhizomes (frame 33)
   b. bulb (frame 32)
   c. corm (frame 31)
   d. tubers (frames 34 and 35)

6. Rhizomes are probably the most readily recognized as stems because they possess nodes, internodes, buds, leaves, and leaf scars. When one slices a bulb in half one can easily see that it is not a root but is made up of fleshy leaves surrounding a terminal bud attached to a flattened stem. A tuber develops as part of a rhizome and possesses "eyes" which are clusters of buds. A corm has attached buds and a microscopic examination of its tissues would reveal that the vascular tissue is arranged in bundles as is typical of stems. (frames 31-36)

CHAPTER SIX
Secondary Growth

1. The two previous chapters have dealt with the origin and structure of the root and stem which together comprise the axis of the plant. You learned that the axis originates from two meristems located at the apices of the axis or, specifically, at the root tip and stem tip.

   What general term can be applied to meristems located at the tips of the plant axis? ______________________________

   - - - - - - - - - - - - - - - - -

   apical meristems

2. Since the first tissues which form the root and stem are produced by apical meristems, we can refer to the root and stem as the primary root and the primary stem. Do apical meristems increase the length of the axis or the diameter of the axis? ______________________________

   - - - - - - - - - - - - - - - - -

   Apical meristems increase the length of the axis.

3. Apical growth constitutes the chief method of growth for most annuals. Growth solely from apical meristems in perennial plants, however, is inadequate, for increase in diameter is necessary to provide support to the increasingly elongated plant axis. Another meristem, the lateral meristem, must, therefore, operate to increase the girth of the plant by the production of extra tissues as discussed in Chapter 1, frame 15.

   What is another name for a lateral meristem? ____________________

   - - - - - - - - - - - - - - - - -

   cambium

4. The purpose of this chapter is to discuss the activity of lateral meristems as they occur in the root and in the stem. Let us, therefore, consider

the structure of the root in this light. In Chapter 4, you learned that the center of the stele of a dicot root is occupied by the xylem. Since this xylem is formed by the (apical/lateral) ________________ meristem, it may be referred to as (primary/secondary) ________________ xylem.

- - - - - - - - - - - - - - - - - -

apical; primary

5. The stele of a dicot root is composed chiefly of vascular tissue. Where is the primary phloem located in relation to the primary xylem?

____

- - - - - - - - - - - - - - - - - -

Circular patches of primary phloem lie between the radiating arms of the primary xylem since xylem and phloem lie in an alternating pattern in roots.

6. Although primary phloem lies in close proximity to primary xylem, the tissues are not confluent; that is, they do not directly touch each other. What separates primary xylem from primary phloem? ________________

____

- - - - - - - - - - - - - - - - - -

A layer of procambial cells separates the two tissues. (Chapter 4, frame 26)

7. In many dicot annuals the procambium which separates the two vascular tissues often differentiates into sclerenchyma. In roots which live for several seasons, however, the procambial cells remain undifferentiated to become a vascular cambium which will produce additional vascular tissue (secondary growth) in the areas between primary xylem and primary phloem.

Where does vascular cambium occur in the monocot root?

____

- - - - - - - - - - - - - - - - - -

Monocots in the temperate zones do not possess cambiums. (See Chapter 4, frame 31.)

8. The photograph on the next page depicts the relationship of the newly formed vascular cambium to primary vascular tissues. Although there

are four arcs of vascular cambiums present in this root, only one is shown here. Each vascular cambial arc does not extend around the ends of the arms of the primary xylem to connect the vascular cambium into a complete ring.

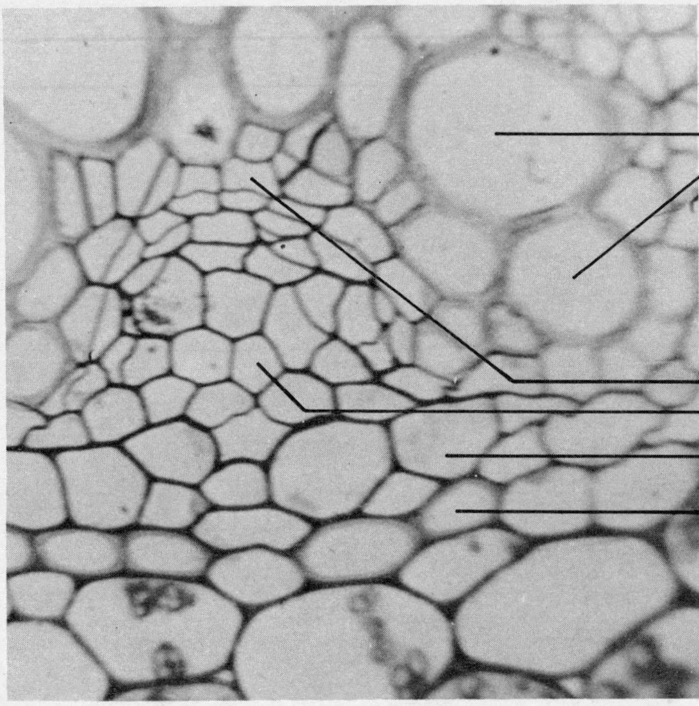

In the photograph above, what tissue prevents the arcs of vascular cambium from closing into a ring? ______________________________

____

- - - - - - - - - - - - - - - - -

The encircling pericycle touches each of the xylem arms, thus preventing the extension of the cambial arcs.

9. In the discussion of roots, you learned that the pericycle reverts to a meristematic condition to form important structures. What are they?

____

- - - - - - - - - - - - - - - - -

Lateral or branch roots are formed from areas of the pericycle.

10. Just as areas of the pericycle become meristematic in branch root production, so do the pericyclic cells at the ends of the xylem arms. They produce strips of vascular cambium which later unite the cambial arcs. Thus, the vascular cambium becomes a closed irregular ring which follows the contour of the primary xylem, as illustrated.

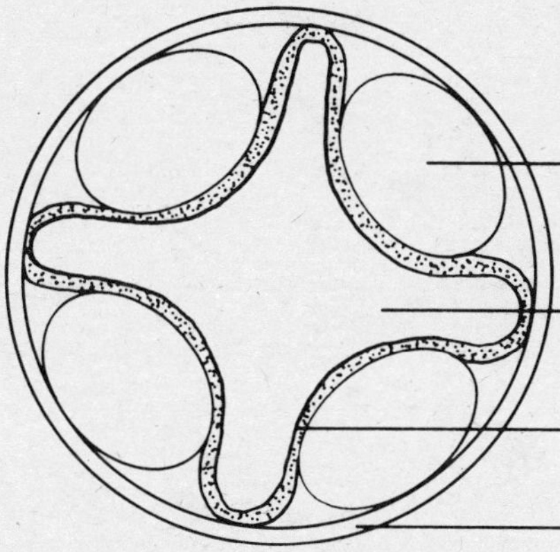

Does the primary phloem lie outside or inside of the vascular cambium? ____________________

- - - - - - - - - - - - - - - - -

The primary phloem lies outside of the ring of vascular cambium.

11. As stated earlier, the vascular cambium forms additional xylem and phloem. Let us see how a cambium operates. In cross section, a vascular cambial cell often appears rectangular and the plane of cell division usually occurs parallel to the axis of the root, as shown in the drawing on the next page.

The drawing reveals that two layers of cells now exist as a result of cell division. Let us assume that the inner layer of cells closest to the primary xylem differentiates into secondary xylem. What do you think will happen to the layer of cells closest to the phloem? ____________________

____

- - - - - - - - - - - - - - - - -

If you guessed that this layer of cells remains meristematic you were correct, because the vascular cambium would disappear after its first division if both layers of cells differentiated into vascular tissue.

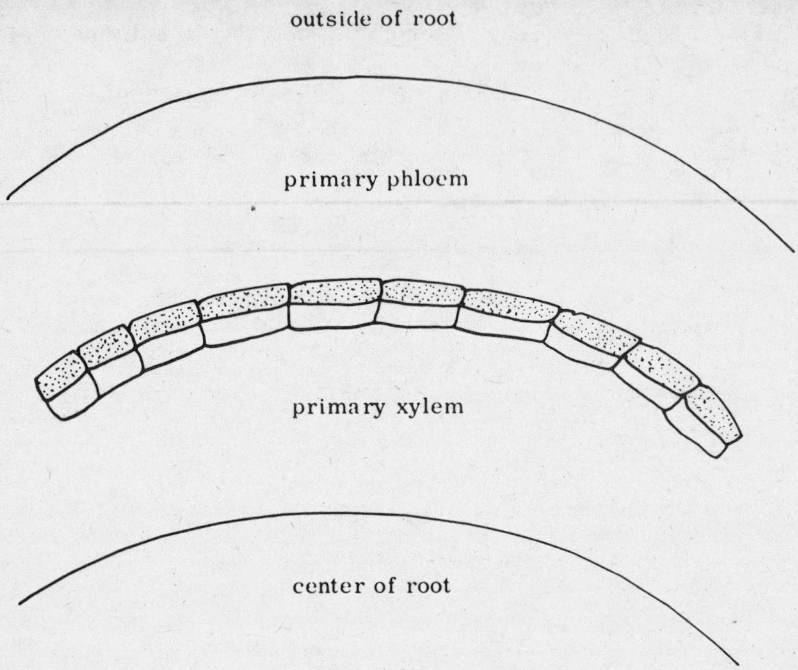

12. In subsequent divisions the vascular cambium will give rise to a layer of secondary phloem cells on the outside of the persistent vascular cambium. Thus, layers of secondary phloem and secondary xylem are increased by the vascular cambium which will remain between them. You must not infer, however, that the vascular cambium produces xylem and phloem in a one-to-one fashion. On the contrary, more secondary xylem is produced by the vascular cambium than secondary phloem. In addition, the thin nature of phloem cell walls (sieve tubes, companion cells, and phloem parenchyma) causes the older phloem tissues to be crushed out of existence as new tissues are formed against them.

There is a greater accumulation of secondary xylem than secondary phloem in a root with secondary growth, for two reasons. These are:

____

____

- - - - - - - - - - - - - - - - - -

More secondary xylem is produced by a vascular cambium than secondary phloem and older phloem is crushed out of existence by newly formed tissues.

13. As secondary growth begins in the root, the portions of the vascular cambium formed by the arcs of procambial cells are more active than the cambium lying at the ends of the xylem arms. The diagram on the next page shows the appearance of a root with initial secondary growth.

Label the location of the vascular cambium in the drawing below.

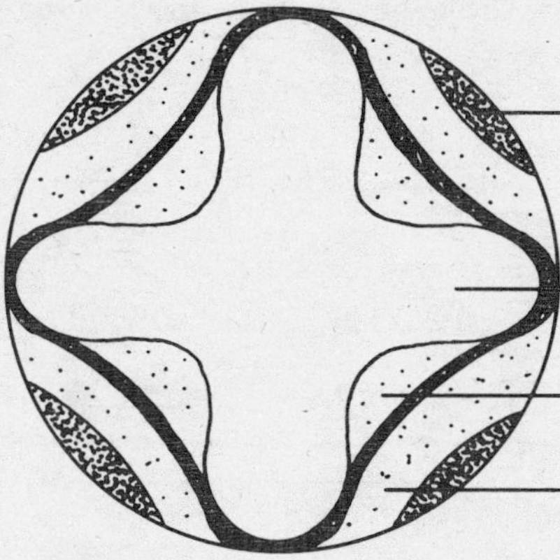

- - - - - - - - - - - - - - - - -

The vascular cambium is always located <u>between</u> secondary xylem and secondary phloem. Since the cambium forms new tissue both externally and internally, it can lie in no other position.

14. Extensive secondary growth serves to build a complete ring of secondary tissues around the primary xylem while the primary phloem is crushed out of existence. Therefore, the stele is eradicated as a discrete region of the root. Label the primary xylem in the photograph below.

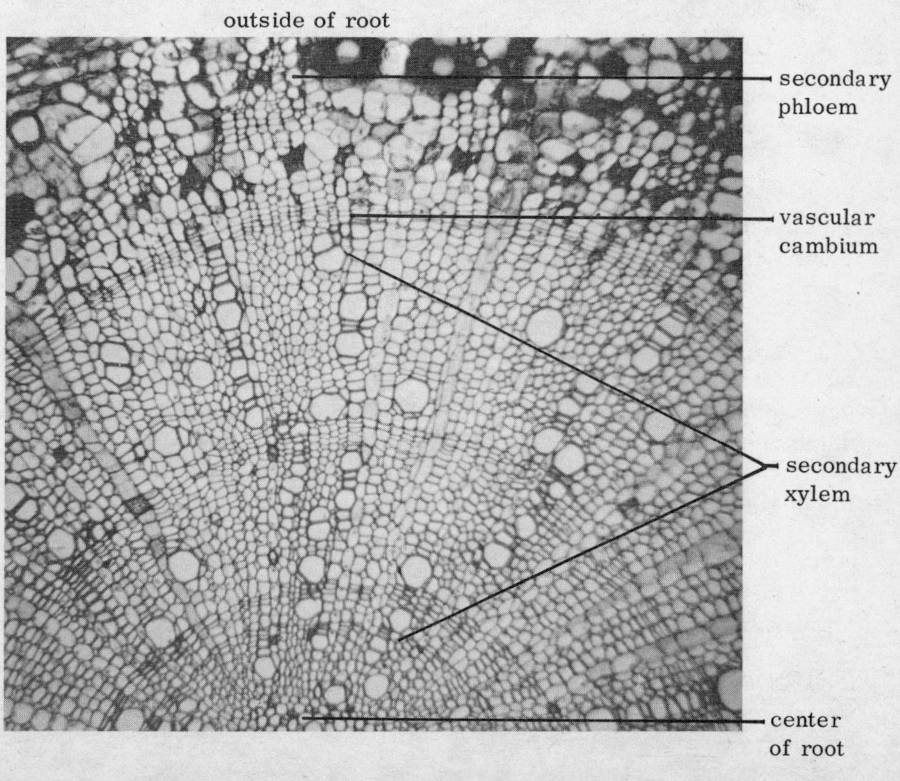

- - - - - - - - - - - - - - - - -

The primary xylem occupies the exact center of the root as it did when it was a component of the stele.

15. We have considered the production of secondary vascular tissue and its effect upon the elimination of the stele, but you must not forget that the stele is surrounded by a broad cortex and the epidermis. Obviously, these two regions cannot exist intact while the circumference of the root is increasing by secondary growth.

    Let us review two facts which you learned earlier. The innermost layer of the cortex is called the ____________________ and the outermost limit of the stele is marked by the ____________________.

- - - - - - - - - - - - - - - - -

endodermis; pericycle

16. You have seen on two previous occasions that the pericycle can become meristematic. Now, as secondary growth takes place in the root, a second lateral meristem may arise from the pericycle at the same time that some pericycle cells close the ring of vascular cambium. The cork cambium formed by the pericycle produces cork cells which replace the cortex and epidermis.

    What three functions does the pericycle perform in a dicot root?

    (1) ____________________

    (2) ____________________

    (3) ____________________

- - - - - - - - - - - - - - - - -

The pericycle forms: (1) lateral or branch roots, (2) a portion of the vascular cambium, and (3) the cork cambium. (any order)

17. The cork cambium (occasionally called a phellogen) functions essentially in the same manner as the vascular cambium. An extensive group of cork cells (phellum) is produced toward the outside of the root while a limited number of parenchyma-like cells (phelloderm) is formed toward the center of the root. The cork cambium together with its derivatives comprise the periderm.

    What three kinds of tissues make up the periderm? ______________

    ____

- - - - - - - - - - - - - - - - -

phellogen, phellum, and phelloderm (any order)

18. The periderm, thus originating from the pericycle, replaces the endodermis as well as the other cortical cells. The epidermis, too, sloughs off as secondary growth from the two lateral meristems proceeds. As the root increases in age, other cork cambiums may originate from cells of the secondary phloem. Thus, the original periderm formed from the activity of the pericycle may be replaced by others of different origin. The following photograph illustrates the appearance of a root with extensive secondary growth.

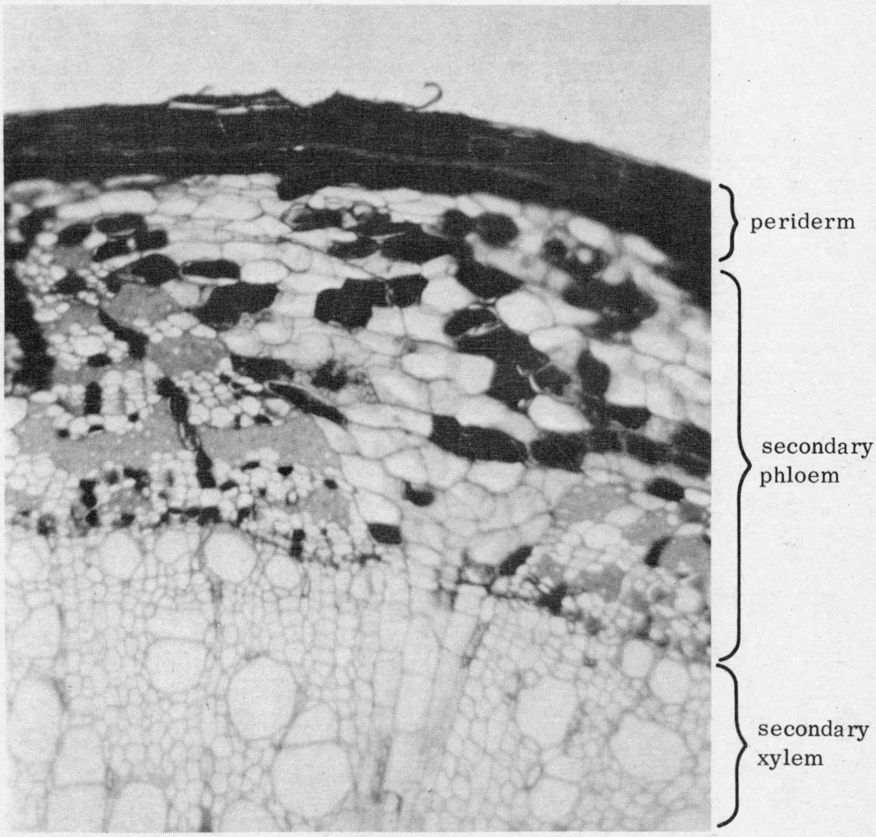

In the above photograph, what secondary tissues lie external to the vascular cambium? ______________________________

____

- - - - - - - - - - - - - - - - -

secondary phloem and the periderm which consists of phelloderm, phellogen, and phellum

19. The secondary tissues formed in the root possess distinctive characteristics worthy of comment. Since an old root closely resembles a stem of the same age, we will consider the aspects of secondary xylem, secondary phloem, and the periderm as we encounter them in the stem where they are more familiar to you.

If the epidermis of the root has been replaced by a corky layer, would you expect a root with secondary growth to absorb much water from the soil? Justify your answer. ______________________

____

____

- - - - - - - - - - - - - - - - -

No. The root hairs of the epidermis are gone and the nearly impermeable nature of the cork tissue will prevent much water penetration.

20. While the root portion of the plant axis increases in diameter by secondary growth, the stem, too, increases its girth by cambial activity. You should recall that the xylem and phloem in a collateral vascular bundle of a young dicot stem are separated by a lateral strip of procambium. Since the vascular bundles are usually separated from each other by pith rays of considerable width, the procambium of a young stem is also broken into isolated lateral bands.

As a review of the internal anatomy of a typical dicot stem, label primary xylem, primary phloem, procambium, and pith ray in the photograph.

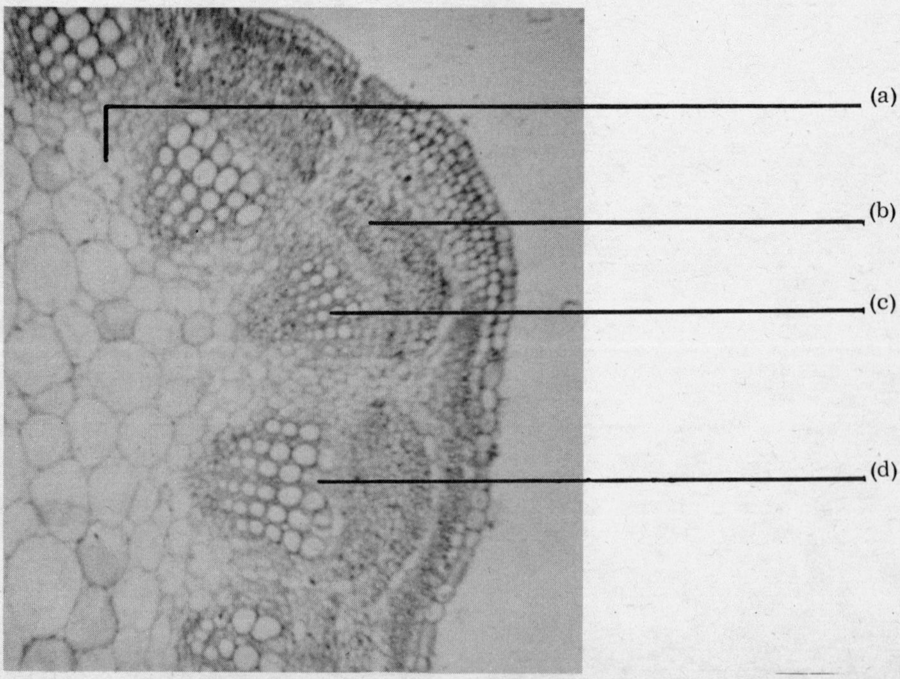

- - - - - - - - - - - - - - - - - -

a. pith ray; b. primary phloem; c. primary xylem; d. procambium

21. As in roots, a stem which lives for only a single season has little or no cambial activity. In perennial plants, however, some of the parenchyma cells in each pith ray revert to a meristematic condition and begin dividing. As a result, the procambiums of the individual vascular bundles are united into a continuous cambial ring (see illustrations).

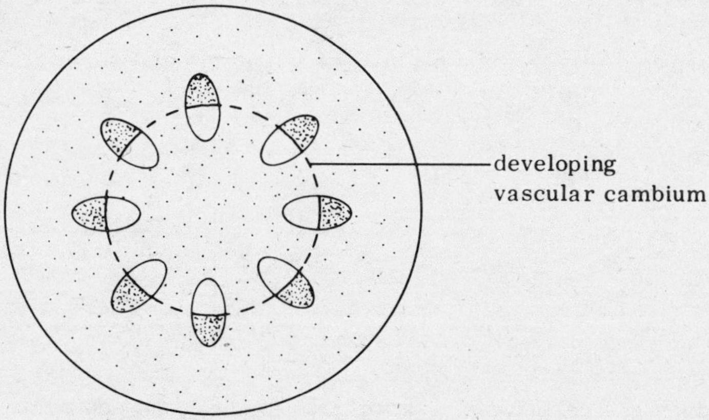

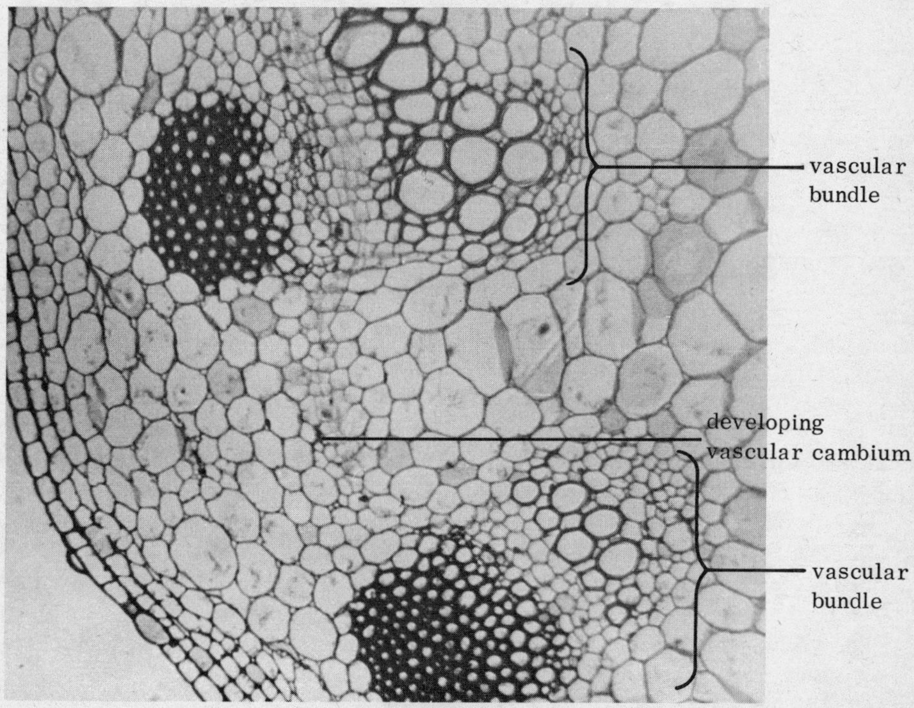

How is the vascular cambial ring closed in the root? ____________

____

- - - - - - - - - - - - - - - - -

The pericycle becomes meristematic in the root and unites the arcs of vascular cambium.

22. Once established, the vascular cambium produces secondary xylem and secondary phloem in the same manner as described in the root. The vascular bundles of the stem disappear as discrete entities as the primary phloem is crushed and the primary xylem is buried by secondary xylem. After a period of extensive secondary growth, older stems and roots closely resemble each other.

What occupies the exact center of a woody stem? ____________

____

What occupies the exact center of a woody root? ____________

____

- - - - - - - - - - - - - - - - -

Parenchyma cells forming a pith are found in the center of a stem. Primary xylem occurs in the center of a root.

23. The primary tissues comprising the cortex and epidermis of the primary stem must be eventually replaced as the diameter of the stem increases due to the activity of the vascular cambium. As in the root, a cork cambium arises in the stem to produce a periderm.

From what tissue does the first cork cambium arise in the root?

____

- - - - - - - - - - - - - - - - -

The cork cambium arises initially from the pericycle of a root.

24. Since the cork cambium arises from the outer stelar tissue in the root, the entire cortex and epidermis are quickly replaced by secondary tissue. In stems, however, the first cork cambium usually forms in superficial cells of the cortex and the periderm thus formed isolates very little cortex in addition to the epidermis. In subsequent years, a succession of cork cambiums arise from successively deeper cortex cells until finally a cork cambium arises from cells in the secondary phloem. Thus, the cortex is eventually and slowly replaced by a number of periderms (see diagram on next page).

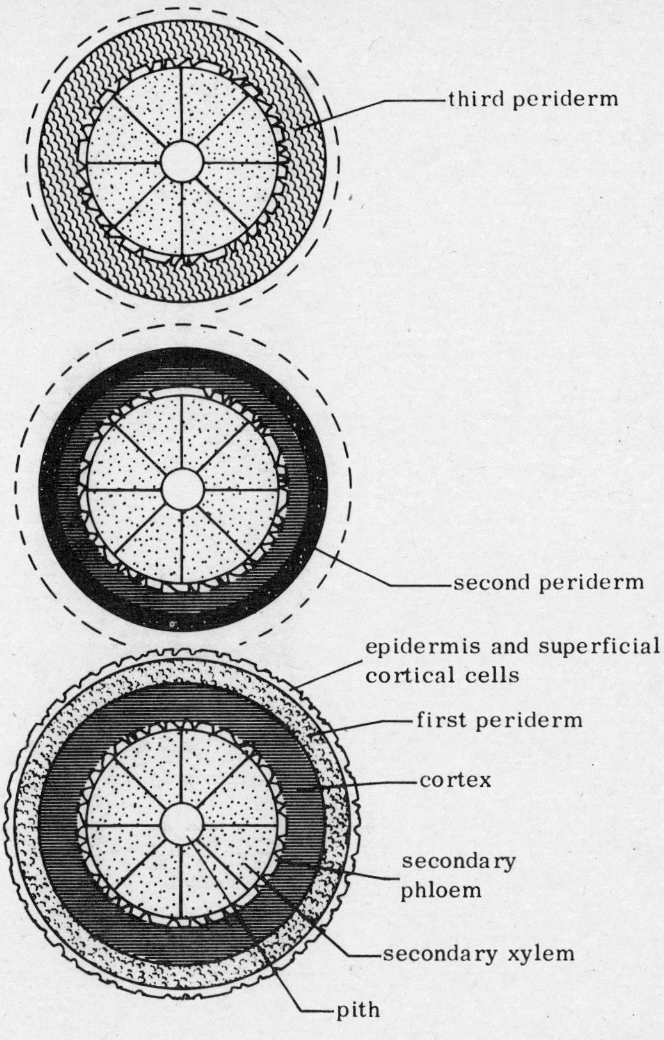

The bark is frequently divided into outer bark and inner bark. If the periderm constitutes the outer bark of a seventy-year-old tree, what tissue comprises the inner bark? ______________________

- - - - - - - - - - - - - - - -

The secondary phloem constitutes the inner bark.

25. On the next page, study the labeled photograph of a basswood stem showing secondary growth and then label the diagram below it, identifying secondary xylem, secondary phloem, cortex, periderm, vascular cambium, and pith.

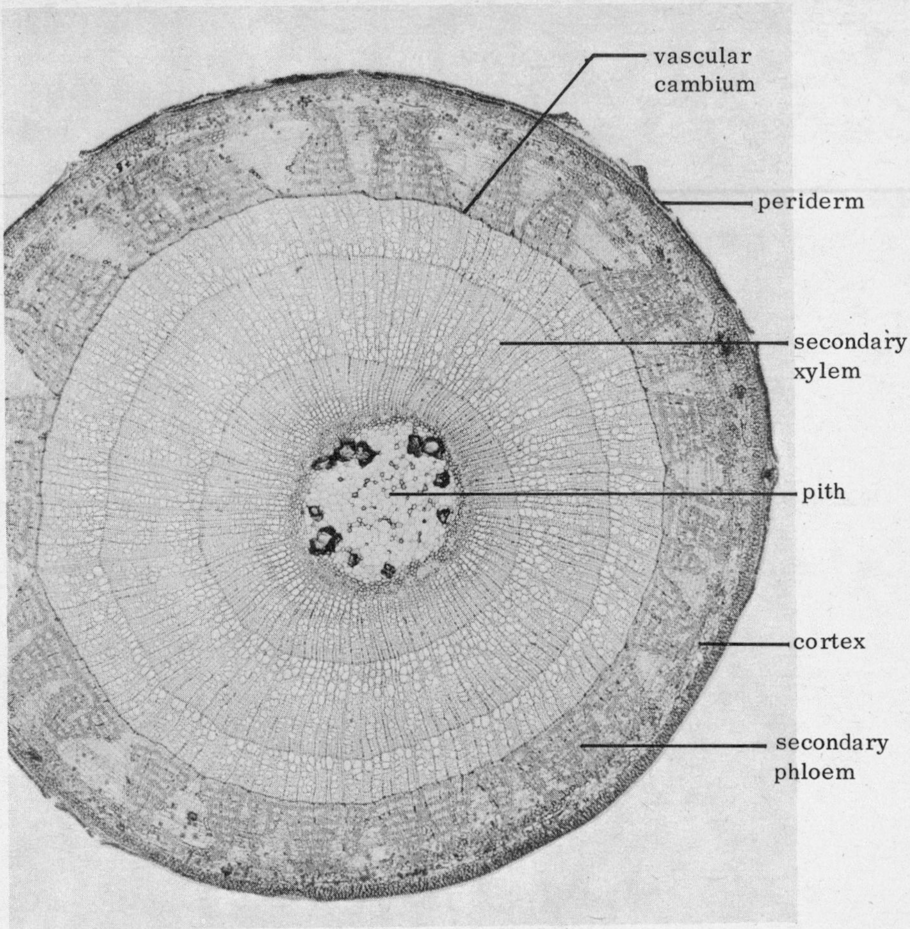

(Courtesy Carolina Biological Supply Company.)

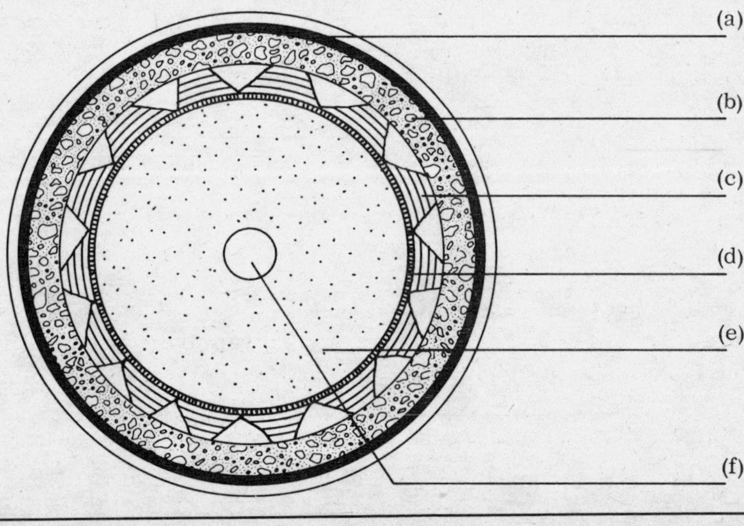

- - - - - - - - - - - - - - - - -

a. periderm; b. cortex; c. secondary phloem; d. vascular cambium; e. secondary xylem; f. pith

26. In a twenty-year-old stem of an elm tree, which member (1 or 2) of each of the following pairs of tissues would you expect to be less abundant?

_______ a. (1) secondary xylem; (2) secondary phloem

_______ b. (1) phelloderm; (2) phellum

_______ c. (1) epidermis; (2) secondary phloem

_______ d. (1) primary xylem; (2) secondary xylem

_______ e. (1) secondary phloem; (2) vascular cambium

- - - - - - - - - - - - - - - - -

a. 2; b. 1; c. 1; d. 1; e. 2

27. Answer each of the following questions by selecting the correct answers from among the numbered items listed. Each question requires two answers.

_______ a. Vascular cambium of a perennial dicot stem arises in what two locations?

_______ b. Cork cambium may arise in what two locations in a perennial dicot stem?

_______ c. Vascular cambium in both roots and stems gives rise to what tissues?

_______ d. Cork cambium gives rise to what two kinds of cells?

1. phellum
2. collateral vascular bundle
3. secondary xylem
4. cortex
5. parenchyma cells of pith rays
6. secondary phloem
7. phelloderm

- - - - - - - - - - - - - - - - -

a. 2 and 5; b. 4 and 6; c. 3 and 6; d. 1 and 7

28. We have traced the development of secondary tissues in the stem, and we have learned that most of the primary tissues have been replaced by tissues which were formed by the cambiums. Let us now consider the nature of secondary xylem as it occurs in both the root and the stem. An extensive accumulation of secondary xylem is referred to as wood. The following discussion, therefore, is concerned with wood anatomy.

Place a check before each of the following statements which is true.

________ a. Apical meristems may produce large amounts of wood under proper conditions.

________ b. Wood may be defined as secondary vascular tissue.

________ c. Wood is produced by a vascular cambium only.

________ d. Wood is composed in part of phloem fibers.

________ e. Wood is secondary xylem.

- - - - - - - - - - - - - - - - - -

c and e are correct statements

29. Almost everyone knows that the age of a tree can be determined by counting the annual growth rings in wood as seen in a cross section of the tree. Have you wondered why each year's accumulation of xylem forms distinct rings in wood? The answer lies in the variation in size of the xylem units. Large and relatively thin-walled xylem cells are formed by the vascular cambium as it begins a period of vigorous growth in the spring. As the season advances, the vascular cambium produces progressively smaller cells with thicker walls until the time that growth ceases in the fall. The cells that were formed last differ greatly in appearance from those formed in early spring. The early and late wood may be called springwood and summerwood, respectively.

How many vascular cambiums are present in the photograph of a woody stem on the next page? (Look at the photograph and write your answer before looking at the answer given here.) ________________

- - - - - - - - - - - - - - - - - -

Only one vascular cambium is present and it lies between the secondary xylem and secondary phloem. A vascular cambium could not function properly in producing both xylem and phloem if it were lying between springwood and summerwood.

30. Does the annual ring produced in the fifth year of a tree's life lie closer to the pith or closer to the secondary phloem in a seventy-year-old tree?

____

\- - - - - - - - - - - - - - - - -

The fifth annual ring lies closer to the pith because secondary growth proceeds outward from the center of the axis. Thus, the oldest ring of xylem lies closest to the pith and the last ring is formed next to the vascular cambium lying adjacent to the secondary phloem.

31. The oldest xylem rings in a tree function as water-conducting tissues for many years, but the cavities in their vessels and tracheids become increasingly restricted in size by the deposition of minerals from the water moving through the xylem. Eventually, the xylem units become permanently clogged and thus nonfunctional.

Clogged xylem becomes dark in color and contrasts sharply with the lighter colored and still functional younger annual rings. The terms heartwood and sapwood are used to describe the position and function of these two areas of wood (see photograph).

Criticize the following statement: Heartwood is dead xylem tissue while sapwood is composed of living cells. ____________________

____

____

____

____

- - - - - - - - - - - - - - - - -

The distinction between heartwood and sapwood should be made on the basis of their containing nonfunctional and functional xylem. Parenchyma cells are often the only living units in secondary xylem and while they may live for many years, they often die long before the sapwood becomes non-functional.

32. Heartwood remains relatively dry and solid for many years, and these qualities together with its dark color produce excellent lumber. The variation in color of redwood, walnut, and red cedar results when boards are cut from both sapwood and heartwood.

How can a hollow tree continue to grow and function? ______________

____

____

- - - - - - - - - - - - - - - - -

A hollow tree results from the decay of the nonfunctional heartwood. Since water is still conducted through the intact sapwood and the phloem of the inner bark still transports food, the tree is able to live and even grow.

33. You have learned that xylem tissue is composed of vessels, tracheids, parenchyma, and sclerenchyma. Secondary xylem or wood is no exception. Unlike primary xylem, however, much of the parenchyma in secondary xylem occurs in radiating rays of elongated cells. These xylem or wood rays vary from one to several cells in width and serve as food storage units and paths for the lateral or radial conduction of food and water. Their appearance is illustrated in a cross section of a woody root or stem.

Wood rays should not be confused with pith rays because wood rays are formed in secondary xylem by the action of a cambium, while pith rays are formed by the apical meristem. Pith rays, therefore, are composed of (primary/secondary) ______________________ tissues.

- - - - - - - - - - - - - - - - -

primary

34. Just as wood rays are produced by the vascular cambium, similar parenchyma cells are simultaneously produced in the secondary phloem to form phloem rays. Because wood rays and phloem rays are formed in vascular tissue only, they are frequently called vascular rays. A single vascular ray in a woody stem may radiate from the pith outward to the cortex or periderm. The portion of the vascular ray lying in the wood constitutes a ________________________ ray while the remainder of the vascular ray which lies in the phloem region is identified as a ________________________ ray.

- - - - - - - - - - - - - - - - - -

wood (xylem) ray; phloem ray

35. The examination of a cross section of wood reveals that a wood ray is limited in width. Such a section, however, does not tell us the vertical dimensions of wood rays. When a tree trunk is stripped of its bark and fed repeatedly through the sawmill to make board lumber, vertical sections of wood are cut and later planed to form the common boards one encounters at the lumber yard. These tangential sections which do not come from the exact center of the tree trunk show the vertical aspects of wood rays, as illustrated below.

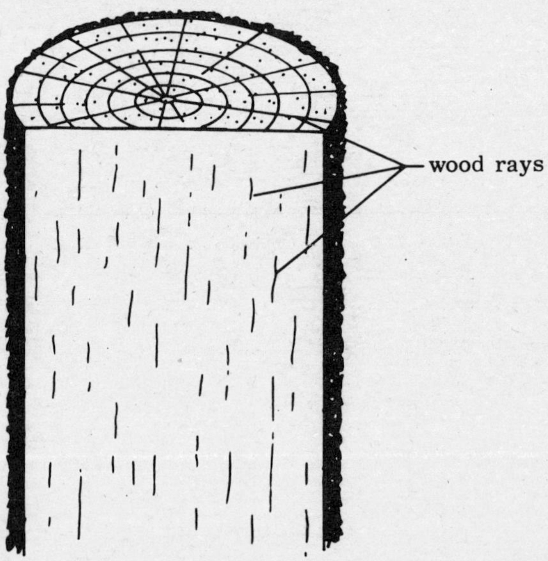

The longitudinal extent of wood rays varies from rays nearly invisible to the unaided eye to those which are nearly two inches long.

An examination of a wood ray cut tangentially should reveal a(n) (end/side) ________________ view of (tracheid/vessel member/wood fiber/wood parenchyma) ______________________________ cells.

- - - - - - - - - - - - - - - - - -

end; wood parenchyma

36. A tangential section of wood not only reveals the nature of wood rays but it also gives a unique view of several annual rings which appear as wavy lines, as illustrated below.

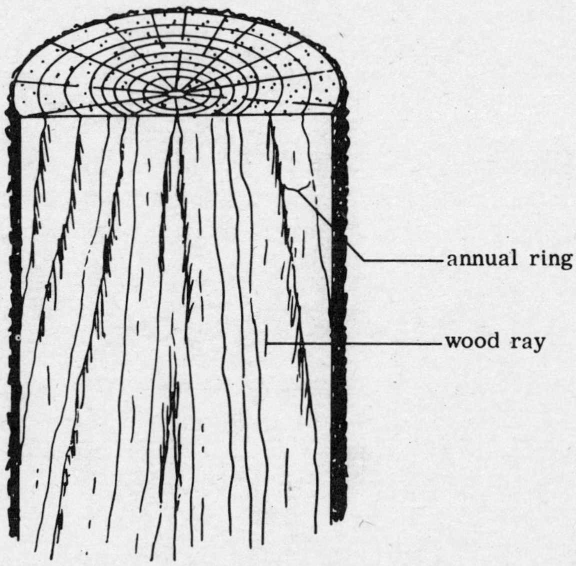

The fact that a tree trunk possesses a greater number of annual rings at its base than it does at its tips, together with a tendency for a tree to produce uneven concentric growth rings explain the wavy, conical nature of the annual rings.

When tangentially sawed boards are used as flooring and subjected to many years of wear, slivers of wood often appear as the result of unequal wear (see diagram).

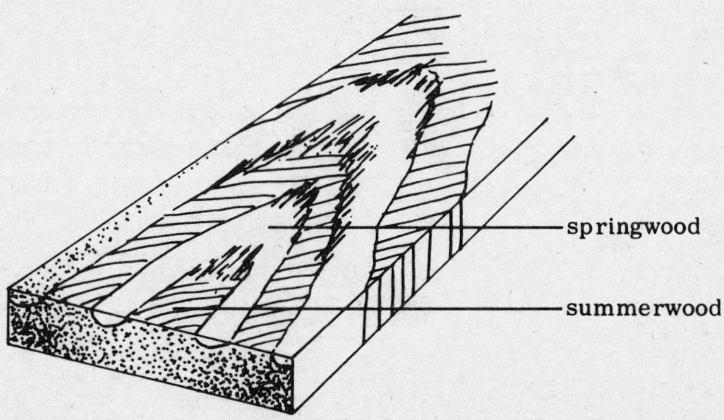

Using your knowledge of the nature of springwood and summerwood, can you guess which part of a particular annual ring forms the sliver and which part has worn away? ____

____

- - - - - - - - - - - - - - - - -

The softer springwood wears away first to leave the harder and denser summerwood as a sliver.

37. If a tree trunk is fed through the sawmill a number of times, eventually a radial section is cut directly through the center of the log. A radial section of wood yields a different view of wood rays and annual rings than that obtained from cross and tangential sections. As the illustration below suggests, annual rings appear as parallel lines and the wood rays may now be seen in lateral view.

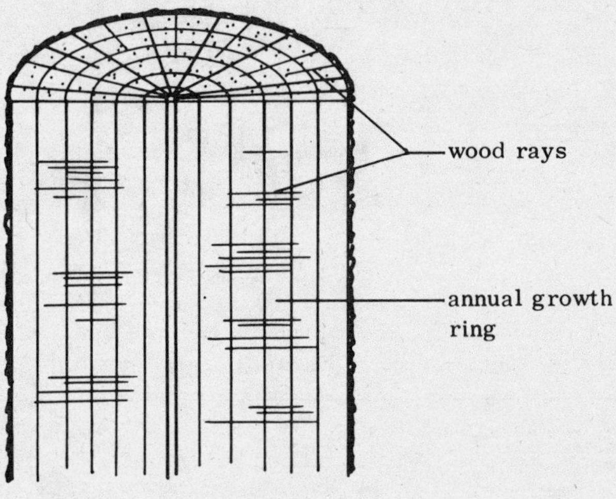

A comparative study of the microscopic nature of wood shows that the xylem components are arranged in a predictable pattern which is unique for each separate species of tree. A wood specialist who has been supplied with thin cross, tangential, and radial sections cut from a piece of wood can easily identify the kind of tree from which the wood specimen was taken.

Match the following words and phrases as a review of the gross structure of wood. Some blanks may require more than one answer.

________ a. Reveals radial nature of wood rays.

________ b. Reveals vertical extent of wood rays.

________ c. Reveals annual rings as wavy lines.

1. tangential section
2. cross section
3. radial section

(continued)

________ d. Reveals concentric nature of annual rings.

________ e. Reveals annual rings as parallel lines.

- - - - - - - - - - - - - - - - -

a. 3; b. 1 and 3; c. 1; d. 2; e. 3

38. We have discussed the appearance of xylem parenchyma as it occurs in radial orientation to form wood rays. Close examination of radial and tangential sections of wood under the microscope, however, reveals that xylem parenchyma also occurs in vertical strands of varying widths and lengths (see photograph). Such vertically aligned cells constitute axial parenchyma. Thus, parenchyma in secondary xylem may be classified as either axial or radial.

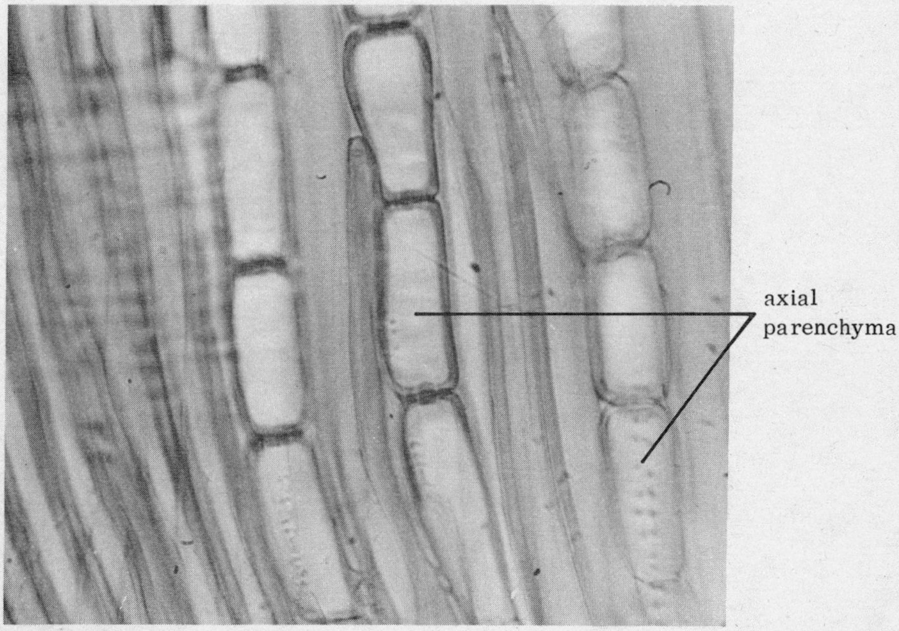

In view of what you know about the function of parenchyma cells, what role do you think axial parenchyma may play in wood? ________________

____

- - - - - - - - - - - - - - - - -

Axial parenchyma, like wood rays, conducts food and water. It may also serve as an important food storage center.

SELF-TEST

Before you proceed to the final chapter of the book, take this Self-Test to aid you in summarizing and reinforcing what you have learned about secondary growth in flowering plants. Questions are to be answered in the light of information in Chapter 6 of this book. Answers are given at the end of this quiz.

1. Select the best answers for each of the following test items. Each blank requires at least two answers.

_______ a. Vascular cambium of a perennial dicot root is derived from:
1. endodermis
2. procambium
3. cork cambium
4. pericycle
5. phellogen

_______ b. Cork cambium of a perennial dicot root may develop from:
1. endodermis
2. pericycle
3. periderm
4. phelloderm
5. secondary phloem

_______ c. Pericycle in a perennial dicot root may develop into:
1. part of the cortex
2. branch roots
3. endodermis
4. vascular cambium
5. phellogen

2. Listed below are a number of tissues or regions encountered in a typical dicot root or stem. Indicate whether each lies inside (I) or outside (O) the vascular cambium.

_______ a. phellogen
_______ b. endodermis
_______ c. primary xylem
_______ d. epidermis
_______ e. primary phloem
_______ f. phellum
_______ g. secondary xylem
_______ h. cortex

3. In a ten-year-old root of an oak tree, which member (1 or 2) of each of the following pairs of tissues would you expect to be <u>more</u> abundant?

_______ a. (1) primary xylem; (2) secondary xylem
_______ b. (1) secondary xylem; (2) secondary phloem
_______ c. (1) phelloderm; (2) phellum
_______ d. (1) secondary phloem; (2) vascular cambium
_______ e. (1) epidermis; (2) secondary phloem

4. Complete the following matching exercise to review some of the concepts presented in this chapter. Some of the blanks may require more than one answer.

________ a. Vascular cambium in both roots and stems gives rise to what tissues?

________ b. Vascular cambium of a perennial dicot stem arises in what two locations?

________ c. Cork cambium gives rise to what two kinds of cells?

________ d. Cork cambium may arise in what two locations in a perennial dicot stem?

________ e. Vascular cambium and cork cambium are both examples of ____.

1. primary xylem
2. epidermis
3. phellum
4. cortex
5. endodermis
6. apical meristems
7. secondary xylem
8. primary tissues
9. phellogen
10. Casparian strips
11. parenchyma cells of pith rays
12. lateral meristems
13. secondary phloem
14. phelloderm
15. primary phloem
16. collateral vascular bundles
17. correct answer not given

5. For each of the following pairs of statements, indicate whether:

1. Both statements I and II are true.
2. Statement I is true but II is false.
3. Statement II is true but I is false.
4. Both statements I and II are false.

________ a. (I) Annual growth rings in a perennial dicot stem are made up entirely of secondary xylem. (II) A new vascular cambium arises each growing season between the springwood and summerwood of the previous season's annual ring.

________ b. (I) Heartwood is made up of summerwood while sapwood is made up of springwood. (II) Annual growth appears as concentric rings when viewed in cross section but not when viewed in a radial section.

________ c. (I) The vertical aspect of a wood ray may be observed in both a tangential and a radial stem section. (II) The lateral aspect of a wood ray may be observed in either a cross section of a stem or in a radial section but not in a tangential section.

6. Secondary xylem is commonly called ____________________. The secondary phloem, remnants of the cortex, and periderm collectively constitute what part of the tree trunk or branch? ____________________

7. Suppose you had three sections (cross, tangential, and radial) of the trunk of a 100-year-old oak tree, all taken at approximately four feet above the soil surface. From which section(s) would you not be able to obtain an accurate measure of the diameter of the tree? ______________________

____

Answers

Compare your answers to the Self-Test with the correct answers given below. If all of your answers are correct, you are ready to go on to the next chapter. If you missed any questions, review the frames indicated in parentheses following the answer before you go on.

1. a. 2 and 4 (frames 7 and 10)
   b. 2 and 5 (frames 16 and 18)
   c. 2, 4, and 5 (frame 16)

2. a. O (frames 16 and 17)
   b. O (frame 15)
   c. I (frame 10)
   d. O (frame 15)
   e. O (frame 10)
   f. O (frames 16 and 17)
   g. I (frame 11)
   h. O (frame 15)

3. a. 2 (frame 14)
   b. 1 (frame 12)
   c. 2 (frame 17)
   d. 1 (frames 11 and 13)
   e. 2 (Chapter 3, frame 4; Chapter 6, frames 24 and 26)

4. a. 7 and 13 (frames 12 and 22)
   b. 11 and 16 (frames 20 and 21)
   c. 3 and 14 (frame 17)
   d. 4 and 13 (frame 24)
   e. 1 (frames 3 and 16)

5. a. 2 (frame 29)
   b. 3 (frames 29, 31, and 37)
   c. 1 (frames 33, 35, and 37)

6. wood; bark (frames 24 and 28)

7. The tangential section would not provide an accurate evaluation of the tree's diameter. (frames 29, 35, and 37)

CHAPTER SEVEN
Leaves, Flowers, and Fruits

1. You have learned how the plant axis is increased in diameter by secondary growth; now let us return to the apical meristem of the stem to review the manner in which leaves arise. In Chapter 5, frame 8, you learned that young leaf primordia arise from the apical meristem simultaneously with the stem tissues. The leaf primordia increase in length and width by continued cell division and enlargement until they become mature leaves.

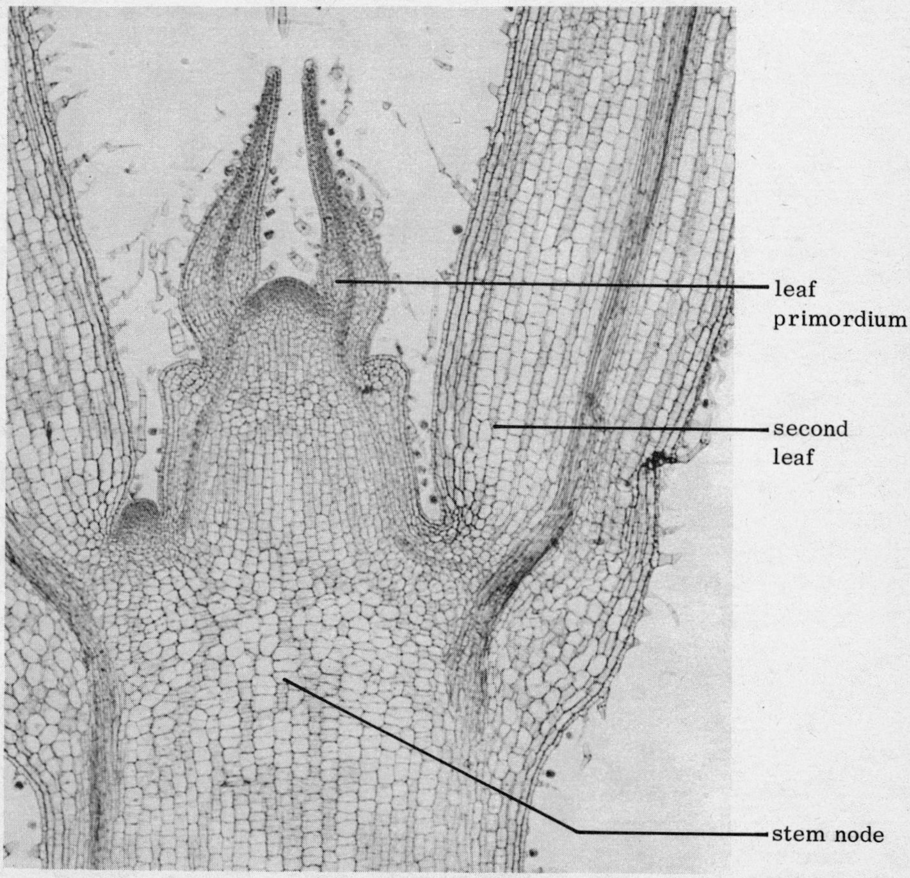

(Courtesy Carolina Biological Supply Company.)

To what part of the stem are the leaves attached? ______________

____

- - - - - - - - - - - - - - - - - -

to the nodes of the stem

2. A typical leaf consists of a thin flat <u>blade</u> which is attached to the stem by a <u>petiole</u> of varying length. In some monocots, such as grasses, the petiole is absent and the blade is attached by a <u>sheath</u> which embraces the stem. Frequently, a pair of <u>stipules</u> occurs at the base of the petiole. Because they vary greatly in form, stipules are often identified solely by their paired nature and by their position at the base of the petiole. When stipules are present, they are considered to be a part of the leaf (see diagram).

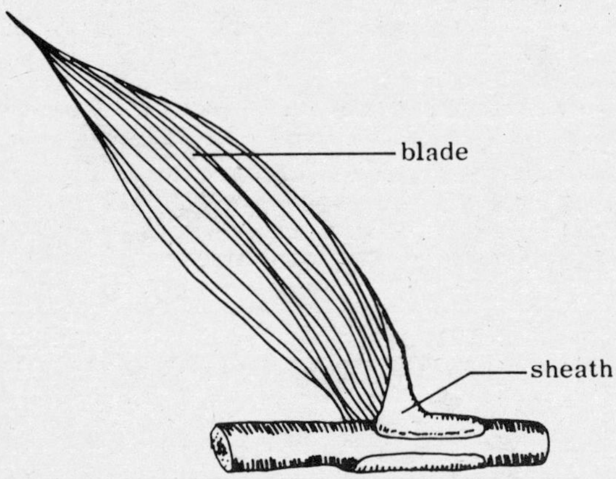

grass leaf

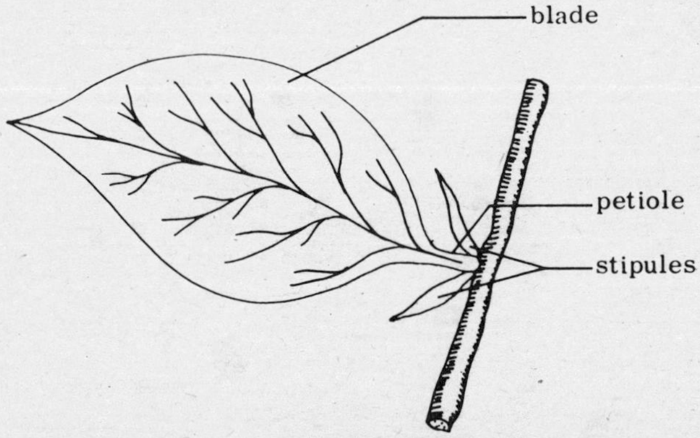

dicot leaf

A black locust tree possesses a pair of sharp spines at the base of each leaf petiole. What can you conclude about the nature of these spines?

The paired spines must be considered stipules because of their position and number.

3. The thin nature of the blade enables the leaf to function efficiently as an organ which manufactures food in the presence of sunlight. Vascular tissue occurs as bundles in the leaf as it does in the stem, and the course of the bundles through the thin blade appears as prominent veins. The pattern set by the branching of the veins constitutes the venation of the leaf. In general, dicot leaves show net venation while parallel venation is characteristic of the vast majority of monocots (see diagram).

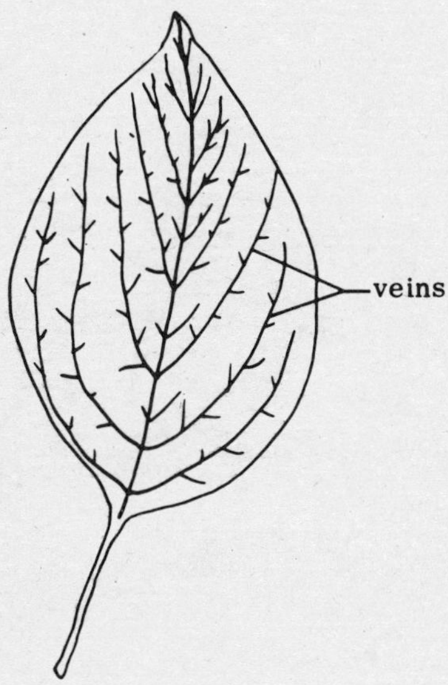

net venation of
a dicot leaf

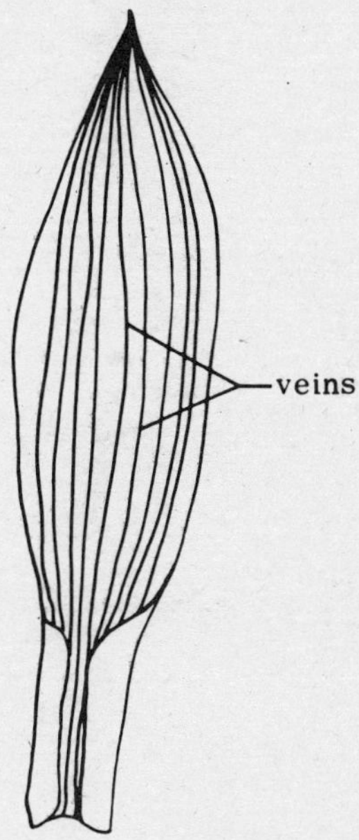

parallel venation of
a monocot leaf

What two characteristics of a grass leaf indicate that grasses are monocots?

---

parallel venation and the presence of a sheath

4. Although a dicot leaf presents a net-veined appearance when the entire network of veins is considered, the main veins in the leaf are arranged in two possible patterns. The largest vein in many leaves extends from the tip of the petiole to the apex of the blade while smaller veins branch off from it at nearly right angles. This featherlike arrangement of the veins is described as pinnate venation. Palmate venation results when several equal-sized veins radiate from the base of the blade much in the same manner as fingers radiate from the palm of the hand. The following diagram illustrates the forms of venation.

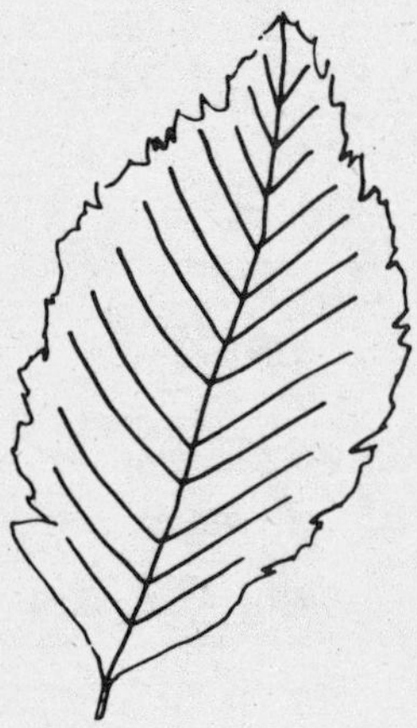

leaf showing
pinnate venation

leaf showing
palmate venation

What two venation patterns may be encountered in a net-veined dicot leaf? ________________________________

- - - - - - - - - - - - - - - - -

pinnate and palmate venation

5. Dicot leaves may vary in form as well as in venation. Simple leaves possess intact and undivided blades such as those of the silver maple and the American elm. The blades of compound leaves, however, are divided into leaflets which resemble miniature leaves. When leaflets are arranged in a featherlike fashion along a central stalk or rachis (rá-kis) as in walnut and ash, the leaves are described as pinnately compound. Buckeye and horse chestnut leaves lack a rachis, for their leaflets are attached to a single point at the tip of the petiole. Their leaves, therefore, are palmately compound (see diagram on next page).

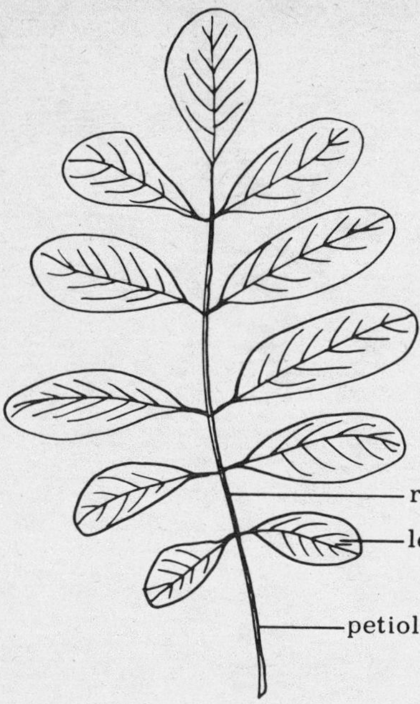

pinnately compound leaf

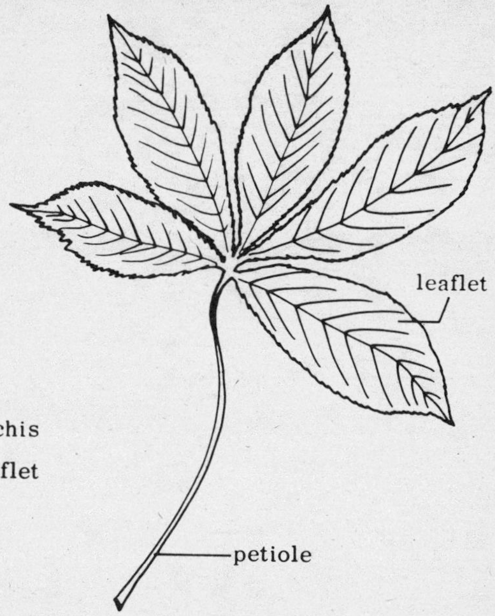

palmately compound leaf

The venation of a compound leaf is described by the venation of any one of its leaflets. Describe the venation of the leaf on the right in the above drawing. ______________________________

____

- - - - - - - - - - - - - - - - -

The palmately compound leaf has pinnate venation.

6. You might ask, "How can I differentiate a pinnately compound leaf from a branch bearing several small but simple leaves?" A petiole of either a simple or compound leaf may be identified by the presence of one or more buds at its base where it joins to the stem. By way of contrast, buds are lacking at the points of attachment of leaflets to the rachis of a pinnately compound leaf.

Which of the two diagrams on the next page illustrates a single pinnately compound leaf? (Look at the diagram and write your answer here before you check the answer below.) ______________

- - - - - - - - - - - - - - - - -

A

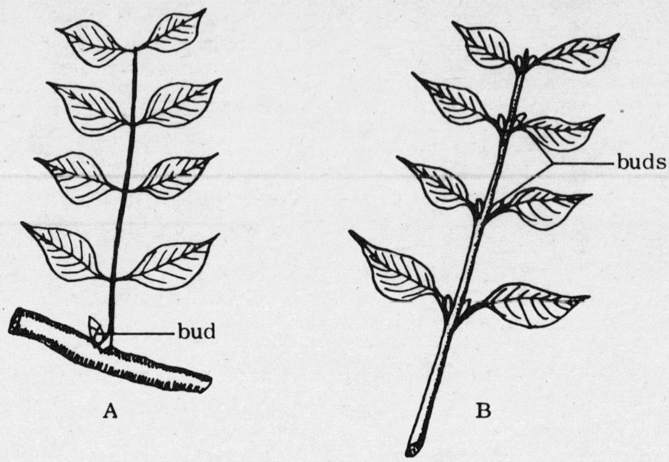

7. Here is a numbered series of drawings of different leaf forms. Match the numbered diagrams to the following brief descriptions.

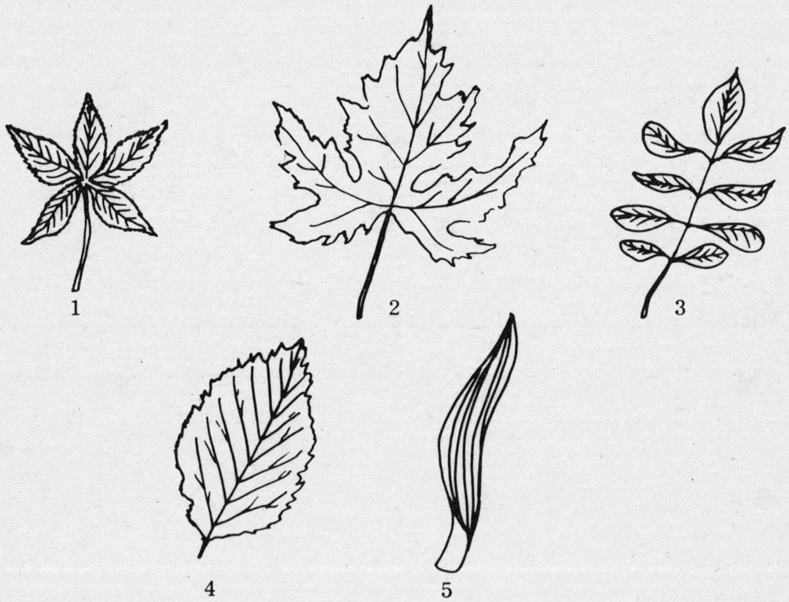

________ a. Typical monocot leaf.

________ b. Compound leaf without a rachis.

________ c. Simple leaf with pinnate venation.

________ d. Pinnately compound leaf.

________ e. Simple leaf with palmate venation.

________ f. Palmately compound leaf.

- - - - - - - - - - - - - - - - -

a. 5; b. 1; c. 4; d. 3; e. 2; f. 1

8. The internal structure of a leaf reflects its function as an organ of food production. Here in the leaf, the important process of photosynthesis occurs, whereby the plant uses the energy of sunlight to combine carbon dioxide and water into sugar. The green pigment, chlorophyll, concentrated in protoplasmic bodies in the cells of the leaf, enables photosynthesis to occur.

Since the green color of tissues is indicative of their ability to photosynthesize, would you expect a mature tree to manufacture food in the winter while its leaves are absent? ____________________

____

- - - - - - - - - - - - - - - - -

Since the trunk and its twigs generally lack chlorophyll, they cannot photosynthesize during the winter, and the plant must live on its food reserves until spring.

9. Examination of a cross section of the blade of a typical dicot leaf reveals a simple tissue organization which allows for efficient photosynthetic activity. A heavily cutinized epidermis covers the upper and lower surfaces of the leaf. Numerous stomata appear in the two epidermal layers in various patterns of distribution. In many leaves, the stomata are confined solely to the lower epidermis, while in others they are equally distributed over both surfaces. In a few water plants, such as the floating leaves of water lily, the stomata are found only in the upper epidermis.

Why would you expect to find an abundance of stomata in the epidermal layers of leaves? ____________________

____

- - - - - - - - - - - - - - - - -

The leaf is an organ of photosynthesis which requires a supply of carbon dioxide from the air to manufacture its food. The stomata provide pores through which air containing carbon dioxide enters the leaves and gaseous waste may leave.

10. The region between the two epidermal layers, commonly called the mesophyll (literally, "middle of the leaf"), is occupied by parenchyma cells, typically of two kinds. A layer of loosely arranged cylinder-shaped cells is located just below the upper epidermis. These palisade cells contain many protoplasmic bodies, the chloroplasts, in which chlorophyll is localized. The remainder of the mesophyll is composed of spongy cells whose loose arrangement suggests their name. Spongy cells possess fewer chloroplasts than the palisade cells.

Using information from the last two frames, label upper epidermis, lower epidermis, palisade cell, and spongy cell in the following photograph.

(a) ____________

(b) ____________

(c) ____________

(d) ____________

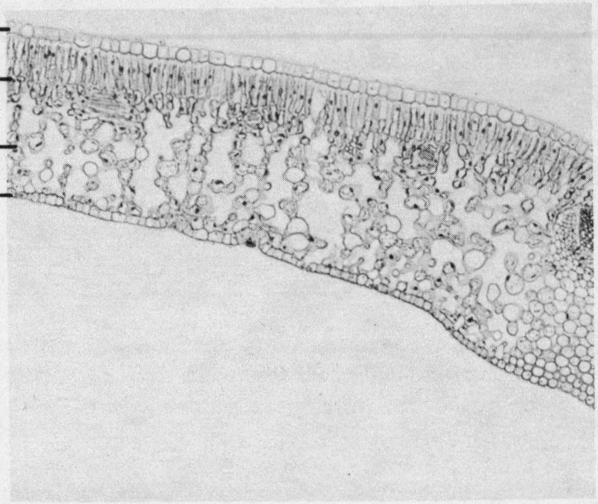

(Courtesy Carolina Biological Supply Company.)

\- - - - - - - - - - - - - - - - - -

a. upper epidermis; b. palisade cell; c. spongy cell; d. lower epidermis

11. The largest veins in a leaf possess tissues characteristic of vascular bundles with the xylem cells lying above the phloem. A poorly developed vascular cambium sometimes produces limited amounts of secondary xylem and phloem. Parenchyma cells surround the large veins and together with collenchyma form prominent ridges on the lower surface of the leaf. The lesser veins run through the spongy layer of the mesophyll and become progressively smaller in diameter until they end blindly in the mesophyll, usually as single tracheids. A single layer of parenchyma cells forms a <u>bundle sheath</u> around each of the smaller veins so that they do not come in direct contact with cells of the mesophyll (see photograph on next page).

    Would you expect the general direction of flow of materials in the phloem of the veins to be out of the leaf or into the leaf? Explain your answer. ____________

\- - - - - - - - - - - - - - - - - -

The direction of flow is out of the leaf since food produced in the leaf is carried in the phloem to other parts of the plant where food is less abundant.

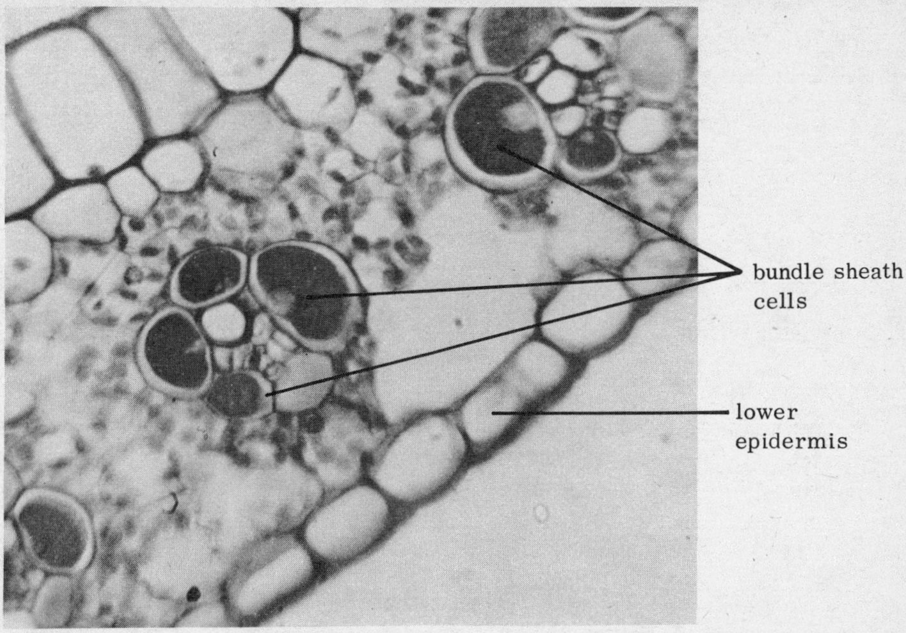

leaf cross section

12. A monocot leaf is exemplified by corn, a member of the grass family. A corn leaf cross section reveals a uniform mesophyll lacking differentiation into palisade and spongy cells. The largest veins are surrounded by an abundance of sclerenchyma fibers while the smaller veins are enclosed by bundle sheaths composed of large parenchyma cells. Greatly enlarged epidermal cells in the upper epidermis of many grasses form bulliform cells which cause the leaf blade to roll in dry weather, conserving moisture.

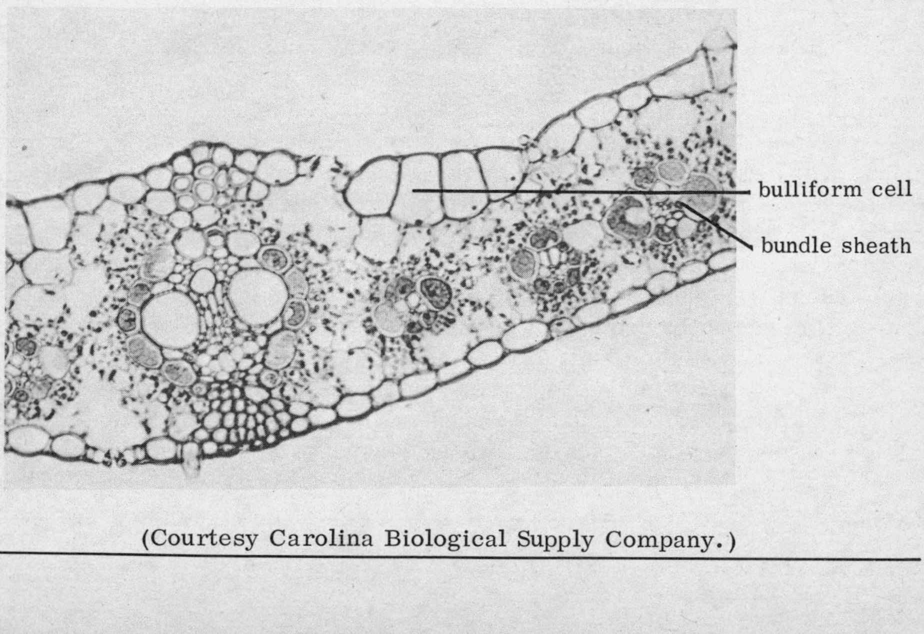

(Courtesy Carolina Biological Supply Company.)

Pili grass (pronounced pee-lee) was used by ancient Hawaiians to fashion their grass huts. What character of the grass leaf would you guess is responsible for its durable nature? ______________________________

____

- - - - - - - - - - - - - - - - -

The fibrous nature of the larger veins gives the grass leaf its toughness.

13. A cross section of a petiole of a dicot leaf reveals one or more vascular bundles embedded in parenchyma which constitutes the chief tissue of this portion of the leaf. Supportive tissue as fibers associated with the veins or collenchyma in strips below the epidermis gives strength to many petioles, as in rhubarb and celery. The number of vascular strands in the petiole varies with the species of plant. In the following diagram of the cross section of a celery petiole, label parenchyma, collenchyma, and vascular bundle.

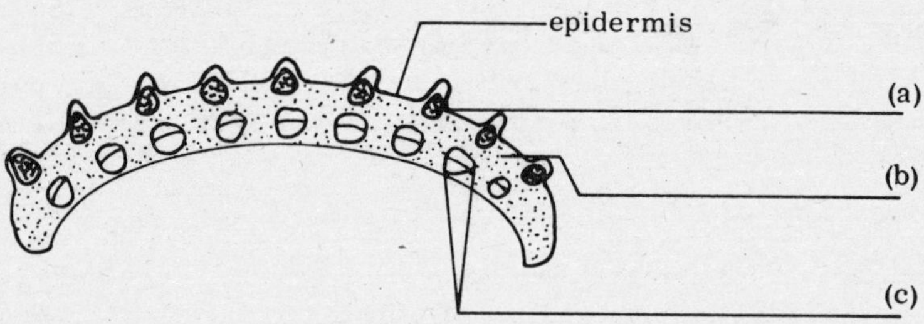

- - - - - - - - - - - - - - - - -

a. collenchyma; b. parenchyma; c. vascular bundle

14. A longitudinal view of a mature leaf petiole enables you to understand how a leaf becomes detached from the twig in the autumn. An abscission (literally, "to cut off") layer develops across the base of the petiole in which the parenchyma cells separate from each other. Since the abscission layer does not involve the vascular strands of the petiole, the leaf remains attached to the stem by the vascular bundles until the wind or the mechanical action of freezing breaks the bundles. A group of cells impregnated with suberin and gum forms a protective layer below the abscission layer while the latter is still active. Thus, the open wound caused by the departure of the leaf is healed by this layer. In woody stems, cork replaces the protective layer and the resulting leaf scar marks the location of the fallen leaf.

The leaves of many oak trees remain on the branches all winter even though they are dead. Can you suggest two possible explanations why the oak trees retain their leaves through the winter? (Write your answer on the next page.)

- - - - - - - - - - - - - - - - -

The abscission layer may not form or the leaves may have tough vascular bundles which resist breaking.

15. The description of the anatomy of a dicot leaf, as presented, is typical of a leaf grown under average moisture conditions, for the size and internal structure of a leaf are greatly affected by the amount of water present in the environment in which it develops. A thick cuticle, a layer of cutinized cells below the epidermis, and a compact mesophyll are often encountered in plants growing in an arid location. These characteristics are in direct contrast to the structure of a leaf formed under wet conditions.

    Can you guess why the thick cuticle, cutinized layer, and compact mesophyll would be advantageous to a plant living under dry conditions?

- - - - - - - - - - - - - - - - -

The features prevent excessive water loss from the leaf.

16. The early part of the life of a flowering plant is concerned with vegetative growth and storage of food, but as it reaches maturity, much of the energy of the plant is utilized in reproduction through flower and seed formation.

    Until the plant matures, the apical meristem produces a leaf bud consisting of a stem of unlimited (indeterminate) growth from which leaf primordia arise as lateral outgrowths. Upon maturity, under the influence of internal physiological changes, the apical meristem forms a stem of limited (determinate) growth with lateral appendages consisting of floral parts.

    Many plants will not produce flowers until they are exposed to an extended period of daylight. Can you guess what physical change occurs in these "long day" plants? ____________________

- - - - - - - - - - - - - - - - -

The apical meristem is induced to form flower buds instead of leaf buds.

17. A typical flower consists of four layers, or whorls, of parts attached to an enlarged determinate stem or receptacle. The lowermost whorl is composed of sepals which, in many instances, are green and leaflike. The petals arise as a whorl immediately above the sepals, followed by a whorl of stamens. The fourth floral part consists of one or more carpels

which become modified into one or more pistils. Since the stamens and carpels are the only parts of the flower necessary to bring about seed formation, they are often called the essential parts of the flower. Sepals and petals, therefore, are the nonessential parts.

Label stamen, sepal, receptacle, petal, and pistil in the following diagram.

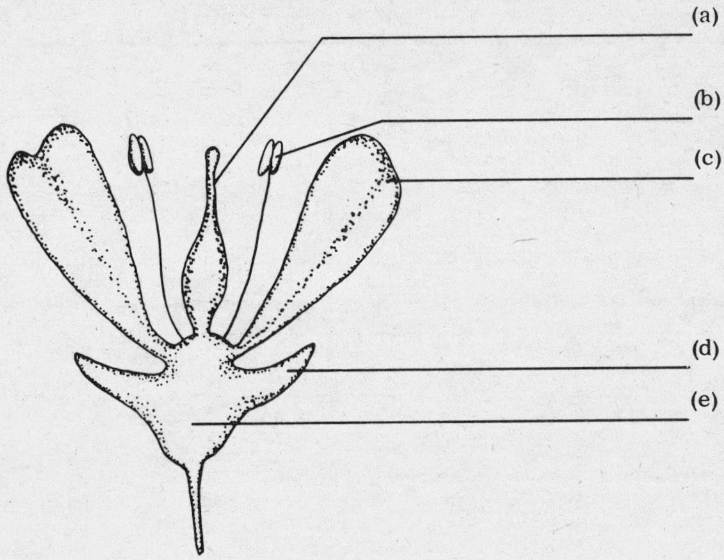

- - - - - - - - - - - - - - - - - -

a. pistil; b. stamen; c. petal; d. sepal; e. receptacle

18. The common origin of flowers and leaves together with the resemblance between the green sepals and leaves, suggest that the floral parts may be modified leaves. The epidermis of sepals commonly bears stomata while the mesophyll is seldom differentiated into palisade and spongy cells. The number of vascular strands going from the vascular bundles of the receptacle into a sepal corresponds to the number of vascular strands entering a leaf of the same plant.

What structures would you expect to find in the mesophyll cells of a green sepal? ______________________________

- - - - - - - - - - - - - - - - - -

chloroplasts

19. The internal structure of the petals is similar to that of the sepals. The epidermal cells of the petals frequently are irregularly shaped and have wavy cell margins. An undifferentiated mesophyll is often colored by pigmented protoplasmic bodies called chromoplasts, and by water soluble

pigments in the watery interior of the cells. A single vascular strand enters almost all dicot petals, while those of monocots often possess more than one.

Petals often fail to develop in some dicot flowers, and the sepals then appear petal-like in shape and pigmentation. In the light of the discussion of the nature of dicot sepals and petals, what anatomical characteristic could you theoretically use to distinguish between the two?

____

____

- - - - - - - - - - - - - - - - -

A single vascular strand is characteristic of most petals and the number of vascular strands in the sepals will be the same as those of the leaves of that plant.

20. A typical stamen is composed of a stalked filament and a two-lobed anther (see diagram).

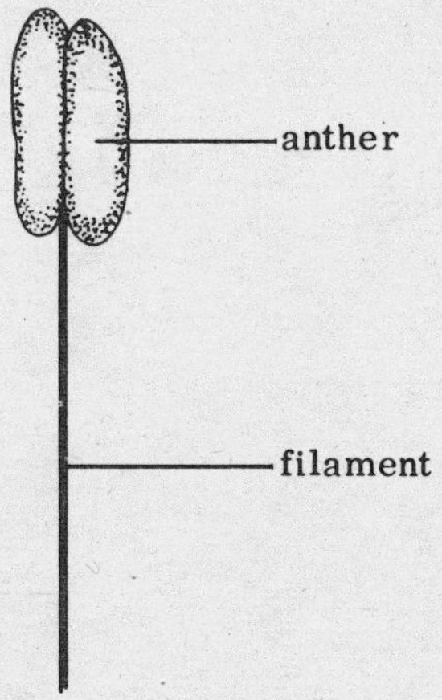

A single vascular bundle typically runs through the filament and ends in the anther. An anther in cross section reveals the important function of the stamens, for special cells present in cavities within the anther develop into pollen grains which are essential to the fertilization of the egg and to the subsequent formation of embryos within the young seeds.

A magnolia flower forms somewhat atypical broad stamens without filaments and with three vascular bundles. In what two ways do these unusual magnolia stamens more closely resemble leaves than do filamentous stamens? ____________________________

____

- - - - - - - - - - - - - - - - -

Magnolia stamens resemble leaves in shape and in the number of vascular bundles, which is greater than the single bundle of filamentous stamens.

21. The carpels occupy the apex of the receptacle of the flower. A carpel is considered to be a leaflike structure which bears ovules or potential seeds along its edges. A midvein (the dorsal bundle) running through the center of the carpel is flanked on each side by the ventral bundles. Label the dorsal and ventral bundles in the following diagram.

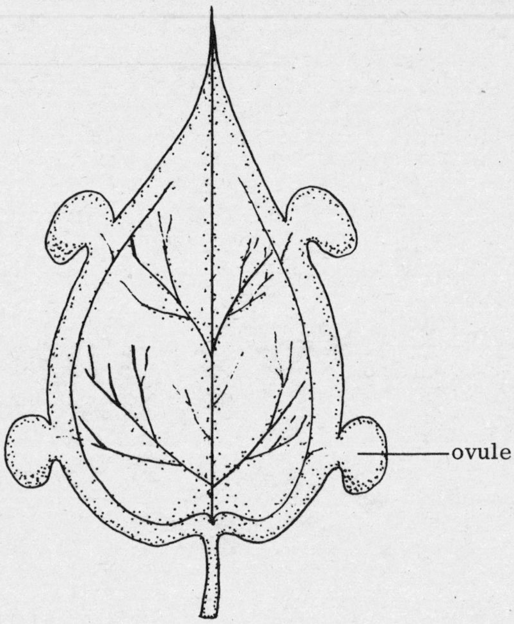

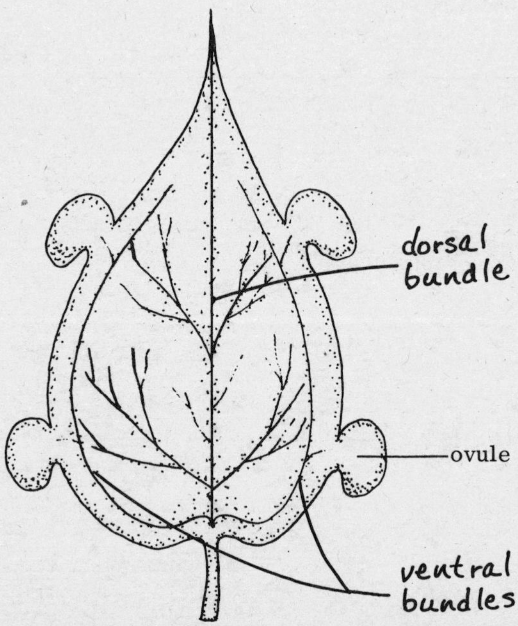

22. Carpels, in reality, do not exist in the form illustrated in frame 21, for each carpel becomes folded in such a manner that its two edges meet and roll inward. Thus, the ovules now lie in two vertical rows within the cavity or <u>locule</u> formed by the enfolded carpel. The following drawings illustrate the external and cross sectional appearance of an enfolded carpel. Again, label the dorsal and ventral bundles as they now appear.

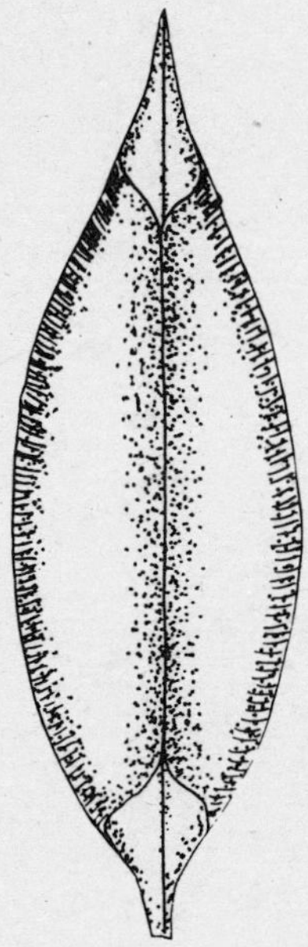

A

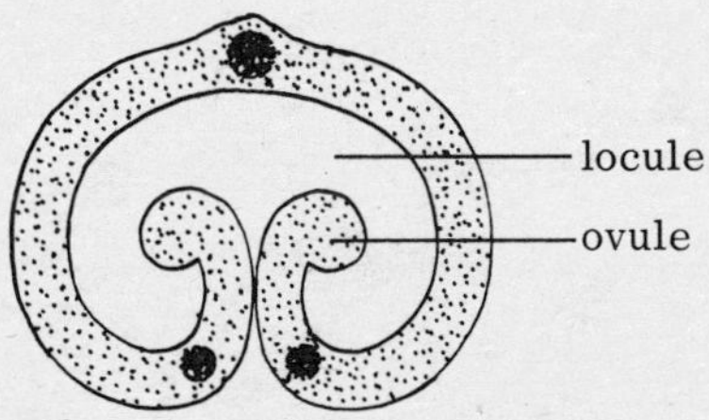

B

Note that the ventral bundles are not visible in drawing A.

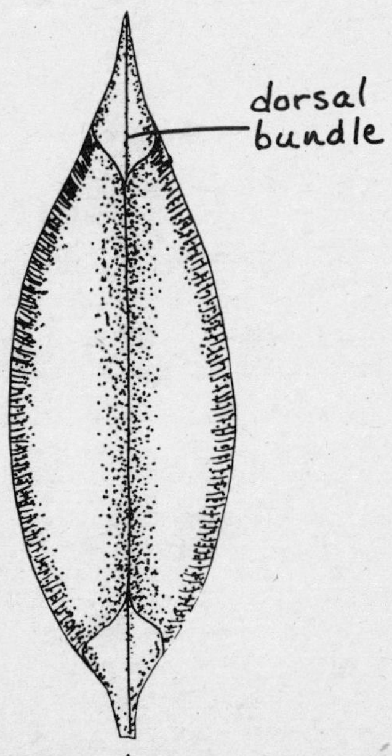

A

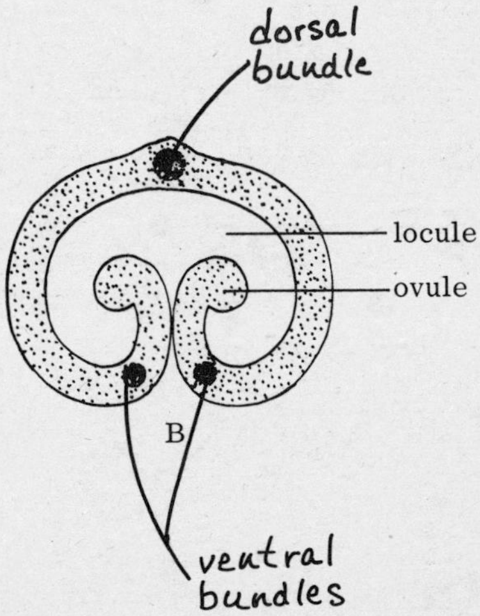

23. Externally, the enfolded carpel becomes modified into a stigma, style, and ovary. A carpel thus differentiated is known as a simple pistil (see diagram).

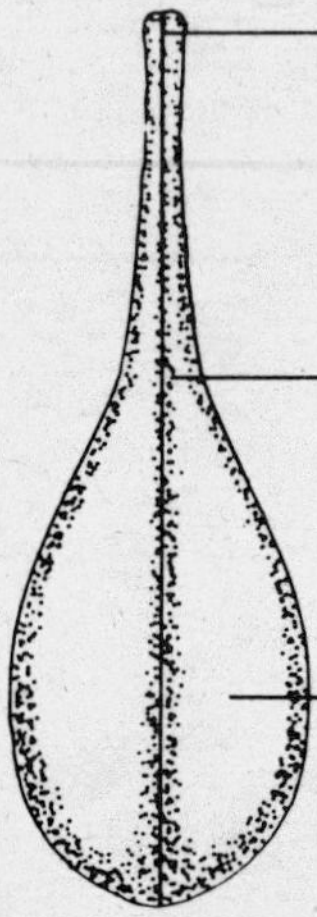

A simple pistil is composed of a single ______________________
which is divided into the following three regions: ______________
____.

- - - - - - - - - - - - - - - - -

carpel; stigma, style, and ovary (any order)

24. Several carpels are often produced by a single flower. Each carpel may develop into a simple pistil so that the flower possesses several pistils. In most instances, however, the carpels fuse in varying degrees to form a single compound pistil. The diagram on the next page illustrates a cross sectional view of an ovary of a compound pistil. Look at the diagram and answer the questions below.

a. How many carpels formed the compound pistil in the diagram?

____

b. How many vascular bundles are present in the illustrated compound pistil? ______________________

- - - - - - - - - - - - - - - - -

a. two; b. six (two dorsal and four ventral bundles)

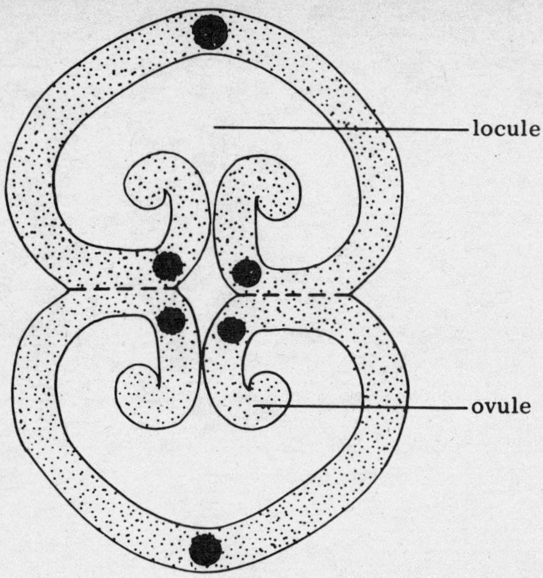

25. The portion of the ovary wall to which the stalked ovules are attached is called the placenta. Hence, an ovary derived from a single carpel possesses a placenta near the union of its folded edges, as illustrated in diagram A below.

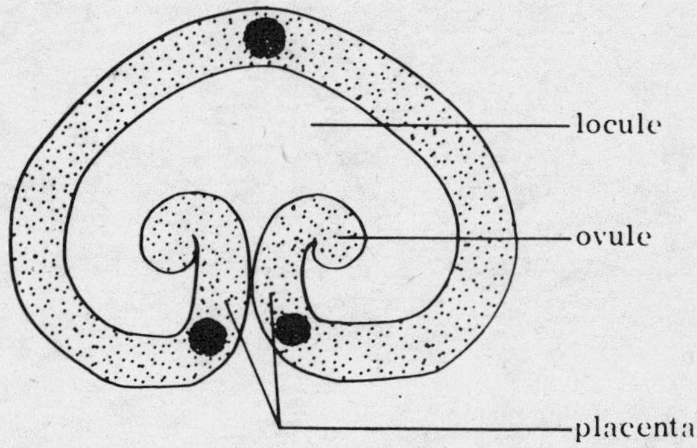

A

The ovaries of compound pistils may show three main types of placentation, depending on the degree of folding and fusing of adjacent carpels. If each carpel of a compound ovary is fused together and the carpels unite in a central fashion, as illustrated in diagram B on the next page, the placentation is said to be axile.

Indicate on diagram B the correct number and position of the dorsal and ventral bundles in the three carpellate ovary.

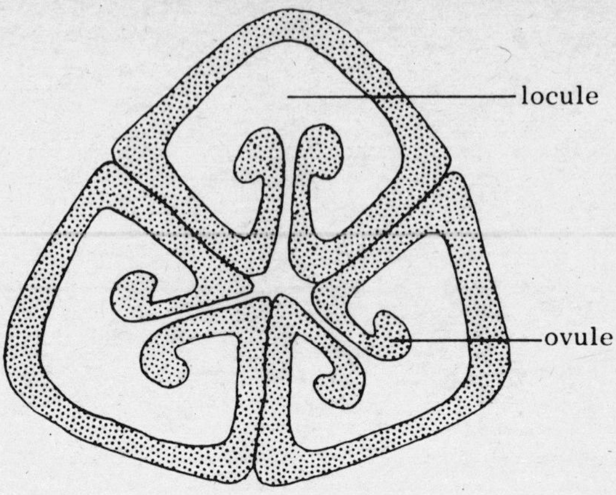

B

- - - - - - - - - - - - - - - - -

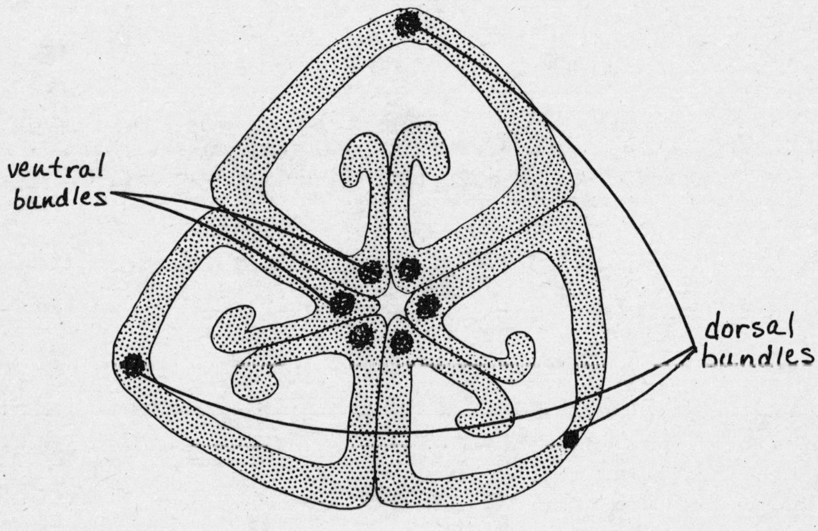

26. If each carpel of a compound ovary remains unfolded and joins adjacent unfolded carpels in the manner illustrated on the next page, the placentation is described as parietal. Furthermore, all simple pistils possess a parietal placentation as shown in figure A of frame 25.

Although they are not all shown, how many vascular bundles should be present in the compound ovary illustrated at the top of the next page?

- - - - - - - - - - - - - - - - -

Nine bundles should be present (three dorsal and six ventral bundles).

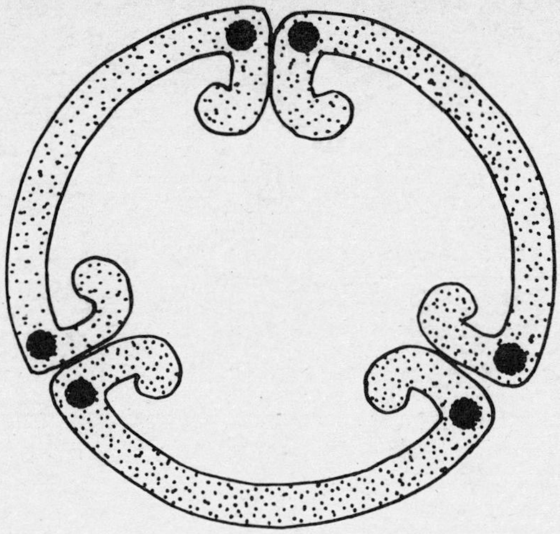

27. Occasionally, the carpel walls which separate the locules of a compound ovary break down upon maturity, or never form. In such instances, the placenta appears to rise into the ovary as a stalked structure from the receptacle. This forms a free-central placentation (see diagram).

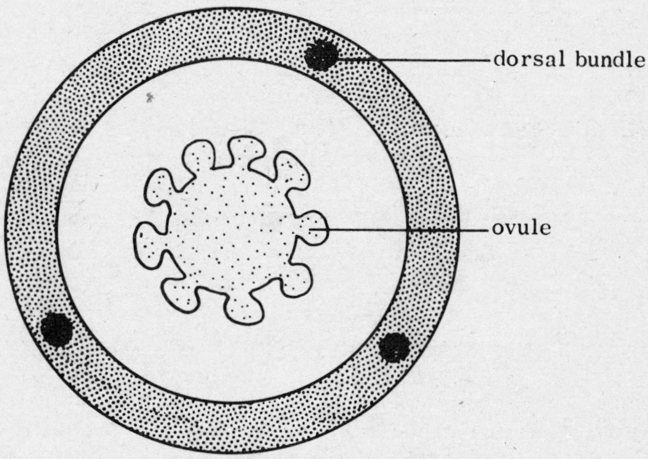

Taking into account the number of dorsal bundles present in the drawing above of the ovary, how many carpels are present? ____________

- - - - - - - - - - - - - - - - -

The presence of three dorsal bundles indicates that there are three carpels present.

28. Identify the types of placentation illustrated in the four diagrams on the next page.

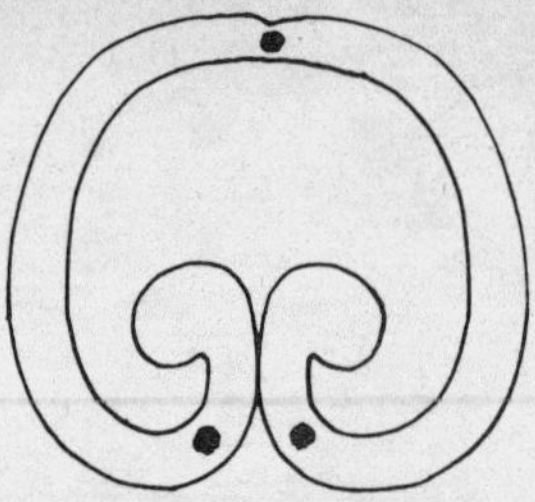

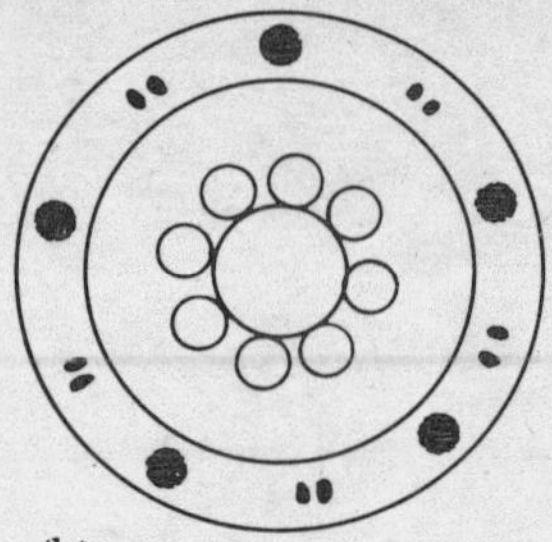

(a)________________ (b)________________

(c)________________ (d)________________

- - - - - - - - - - - - - - - - - -

a. parietal; b. free-central; c. parietal; d. axile

29. In all of the previous illustrations representing placentation, we have assumed that the ventral bundles of the carpels remain isolated and distinct even though carpel walls fuse. Such an assumption is often erroneous, for pairs of vascular bundles may fuse to the extent that it is difficult or impossible to ascertain the nature of a bundle except by deduction. Fusion may involve two ventral bundles from a single carpel or two ventral bundles derived from two adjacent carpels. The illustration on the next page depicts the bundle arrangement of a three carpellate ovary. Look at the illustration and answer the questions below.

    a. How many vascular bundles are illustrated? ________________

    b. How many vascular bundles will be present if adjacent bundles fuse in pairs? ________________

- - - - - - - - - - - - - - - - - -

a. nine (three dorsal and six ventral bundles); b. six (three dorsal and three fused ventral bundles)

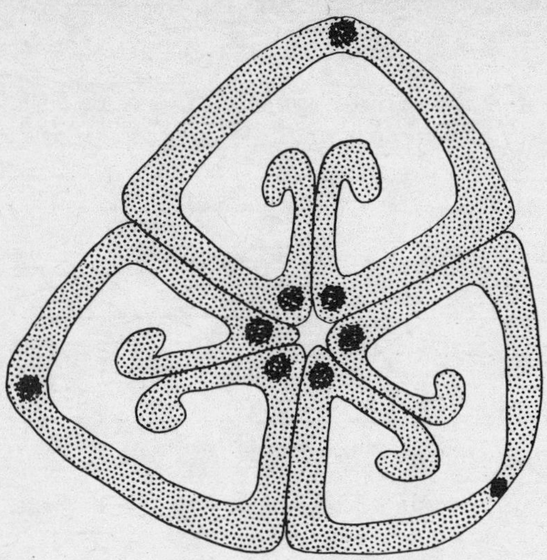

30. As the result of sexual reproduction in the flower, almost all of the ovules form embryos and become seeds. As the seeds mature, the ovary portion of the pistil enlarges and differentiates into a protective structure which encloses the seeds. In the narrowest sense, the mature ovary constitutes a fruit, but in many instances, other floral parts may surround the ovary and participate in fruit formation. In formulating a definition of a fruit, therefore, these other flower parts should be considered.

A fruit, then, is the mature ______________________ plus any other part of the flower which may persist. Since a pea pod contains seeds covered by an ovary, can a pea pod be considered a fruit? ______________

- - - - - - - - - - - - - - - - -

ovary; According to definition, a pea pod is a fruit.

31. The mature ovary wall is properly called the pericarp, and in those instances where no other floral part is involved the term is synonymous with fruit.

Prior to seed formation the young ovary is composed chiefly of parenchyma and vascular tissue. As the fruit begins to mature, the number of cells in the ovary wall usually increases. The pericarp may remain essentially parenchymatous, or various kinds of sclerenchyma cells may also develop to a degree that the mature pericarp is totally different in cellular composition from the young ovary.

The tomato represents a type of fruit. Can you guess what kind of tissue comprises most of the tomato fruit? ______________________

- - - - - - - - - - - - - - - - -

The tomato is composed mostly of thin-walled parenchyma.

32. The pericarp can frequently be divided into an exocarp or outer region, a mesocarp or middle region, and an endocarp or inner region. The cellular composition of the three regions often varies greatly.

A peach is an excellent example of a pericarp which is divided into three regions. The skin represents the exocarp, the "flesh" comprises the mesocarp, and the stone marks the endocarp. Can you guess what type of tissue forms the main portion of the endocarp? ____________

____

- - - - - - - - - - - - - - - - -

The hard stone suggests sclerenchyma, either as sclereids or fibers.

33. A mature fruit may be classified as fleshy or dry depending on its cellular composition. A fleshy fruit is obviously composed of parenchyma which contains much water and soluble constituents. If the pericarp becomes fleshy throughout, the fruit is classed as a berry.

What one characteristic is shared by tomatoes, grapes, oranges, bananas, and green peppers? ____________________

____

- - - - - - - - - - - - - - - - -

They all have a pericarp which is entirely fleshy. They are all berries.

34. Two other kinds of fleshy fruits are drupes, whose endocarps are hard and stony, and pomes, in which a fleshy receptacle surrounds a somewhat dry and papery pericarp.

a. The core of an apple represents the pericarp. The apple, therefore, is an example of what kind of fruit? ______________

b. Which of the following fruits do you think is also an example of a pome: apricot, peach, pear, or watermelon? ______________

c. All stone fruits are examples of ____________________.

- - - - - - - - - - - - - - - - -

a. a pome; b. pear; c. drupes

35. The pericarp of dry fruits is commonly divided into exocarp, mesocarp, and endocarp of varying nature. As you may guess, more sclerenchyma is present in dry fruits than in fleshy ones, and parenchyma, although present in dry fruits, contains less water and is therefore less succulent.

Dry fruits are classed as dehiscent if the pericarp breaks open at maturity and indehiscent if the pericarp remains closed.

In the autumn you may have observed milkweed pods splitting open and releasing their winged seeds to be scattered by the wind. A milkweed pod is an example of a (dry/fleshy) ________________, (dehiscent/ indehiscent) ________________ fruit.

- - - - - - - - - - - - - - - - -

dry, dehiscent

36. Dry, indehiscent fruits are often one-seeded and vary in the nature of their pericarps. A well known kind of dry indehiscent fruit is a grain, in which its single seed is fused to the pericarp. An achene (or akene) is another one-seeded indehiscent fruit in which the pericarp is separate from the seed. A nut is an example of an achene with a hard sclerenchymatous pericarp.

Can you guess what kind of dry indehiscent fruits are represented by each of the following?

________________ a. A sunflower "seed" can be cracked open and its contents eaten.

________________ b. An acorn can be broken open with difficulty to find a "kernel" inside.

________________ c. A "kernel" of corn is surrounded by a covering which cannot be separated from the seed within.

- - - - - - - - - - - - - - - - -

a. The sunflower seed is an achene.
b. The hard outer layer of the acorn indicates that it is a nut.
c. The corn "kernel" is a grain with the single seed fused to the external pericarp.

SELF-TEST

Review and reinforce what you have learned about leaves, flowers, and fruits by taking this short Self-Test. Questions are to be answered in the light of information presented in Chapter 7 of this book. Answers are given at the end of the quiz.

1. Shown below is a series of numbered diagrams of leaves. Match the appropriate diagram with the correct descriptive statement.

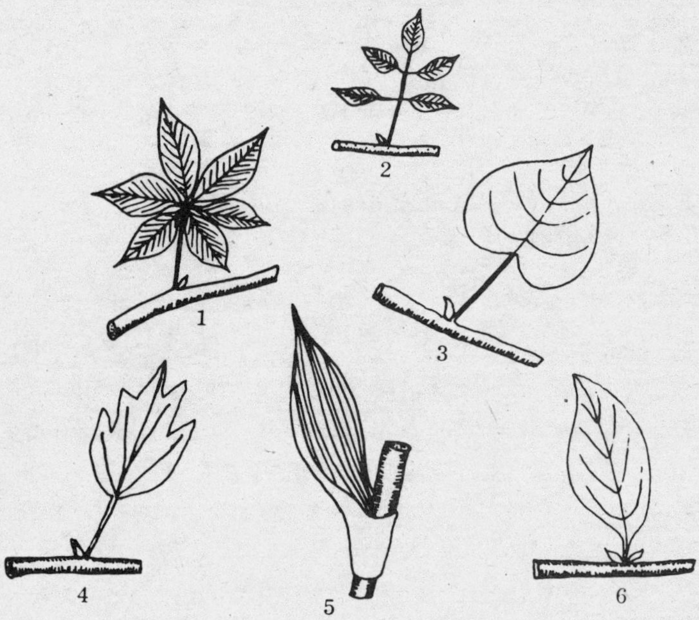

________ a. Palmate venation.

________ b. Pinnately compound.

________ c. Monocot leaf.

________ d. Palmately compound.

________ e. Leaf with stipules.

2. In the following questions select your answer from among the numbered diagrams shown in question 1 and answer by number only. Some of the blanks require more than one answer.

________ a. Which leaves have a rachis?

________ b. Which leaves have a petiole?

________ c. Which leaves are pinnately veined?

________ d. Which leaves have leaflets?

3. Match a numbered leaf part with each lettered description.

________ a.	Cylindrical cells just below upper epidermis.	1.	epidermis
		2.	xylem
________ b.	Leaf cells containing abundant chloroplasts constituting the middle of the leaf.	3.	spongy
		4.	petiole
		5.	parenchyma
		6.	cuticle
________ c.	Cell layer containing stomata.	7.	collenchyma
________ d.	Cells constituting a bundle sheath.	8.	palisade
		9.	phloem
________ e.	A leaf vein is made up principally of these two kinds of tissue.	10.	mesophyll

4. Match two numbered items with each lettered item. One numbered item should be the name given to the kind of fruit described while the other should be a specific example of the kind of fruit described.

________ a. A dry indehiscent fruit containing a single seed fused to the pericarp.

________ b. A stone fruit.

________ c. An achene with a hard sclerenchymatous pericarp.

________ d. A fruit whose pericarp is fleshy throughout.

________ e. A dry one-seeded indehiscent fruit in which the pericarp is separate from the seed.

________ f. A fruit consisting of a dry papery pericarp surrounded by a fleshy receptacle.

Term (name)
1. achene
2. berry
3. drupe
4. grain
5. nut
6. pome

Example
7. acorn
8. cherry
9. corn kernel
10. pear
11. sunflower "seed"
12. tomato

5. In the diagram on the next page label anther, filament, ovary, petal, pistil, receptacle, sepal, stamen, stigma, and style.

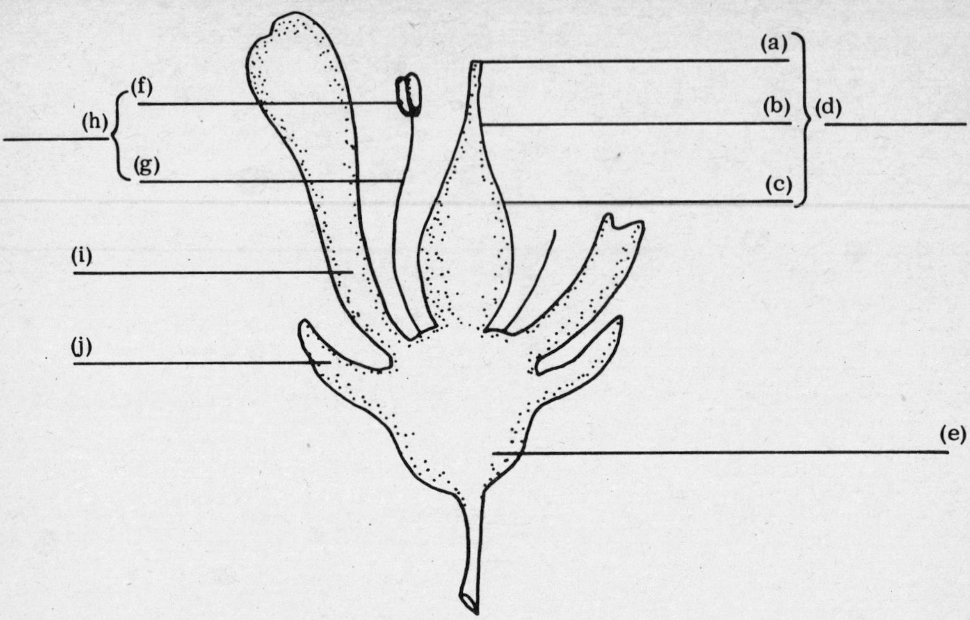

6. Use the diagram in question 5 to answer the following questions.

_______ a. Of the four major flower parts (d, h, i, and j in the diagram), which may be regarded as essential to seed production?

_______ b. As specifically as possible, identify the part of the flower in which the seeds will be produced.

_______ c. As specifically as possible, identify the part of the flower in which pollen grains will be produced.

_______ d. Which of the flower parts is usually most leaflike in appearance?

_______ e. As specifically as possible, identify the part of the flower that will mature into the fruit.

7. Shown on the next page is a cross section of the ovary of a flower. Refer to the diagram to answer the following questions.

a. Is the ovary shown a part of a simple or compound pistil?

____

b. How many carpels make up the pistil of this flower? _______

c. How many dorsal bundles are present in this ovary? _______

d. How many ventral bundles? _______

e. What type of placentation occurs in this ovary—axile, free-central, or parietal? _______________

(continued)

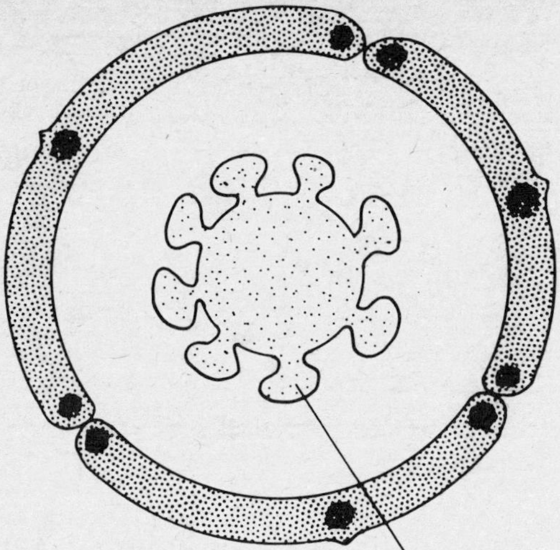

f. What name is given to the object labeled A in the diagram?

____

g. Into what will the object labeled A develop? ____________________

Answers

Compare your answers to the Self-Test with the correct answers given below. If all of your answers are correct, you are ready to go on to the Final Test. If you missed any questions, review the frames indicated in parentheses following the answer before you take the Final Test.

1. a. 4 (frame 4)
   b. 2 (frame 5)
   c. 5 (frames 2 and 3)
   d. 1 (frame 5)
   e. 6 (frame 2)

2. a. 2 (frame 5)
   b. 1, 2, 3, 4, and 6 (frame 2)
   c. 1, 2, 3, and 6 (frame 4)
   d. 1 and 2 (frame 5)

3. a. 8 (frame 10)
   b. 3 and 8 or 10 (frame 10)
   c. 1 (frame 9)
   d. 5 (frame 11)
   e. 2 and 9 (frame 11)

4. a. 4, 9 (frame 36)
   b. 3, 8 (frames 32 and 34)
   c. 5, 7 (frame 36)

d. 2, 12 (frame 33)
e. 1, 11 (frame 36)
f. 6, 10 (frame 34)

5. a. stigma (frame 23)
b. style (frame 23)
c. ovary (frame 23)
d. pistil (frames 17 and 23)
e. receptacle (frame 17)
f. anther (frame 20)
g. filament (frame 20)
h. stamen (frames 17 and 20)
i. petal (frame 17)
j. sepal (frame 17)

6. a. d and h (frame 17)
b. c (frames 21 and 25)
c. f (frame 20)
d. j (frame 17)
e. c (frame 30)

7. a. compound (frame 24)
b. three (frames 24 and 27)
c. three (frame 21)
d. six (frame 21)
e. free-central (frame 27)
f. ovule (frames 21 and 25)
g. seed (frame 21)

Final Test

Review and reinforce what you have learned about plant anatomy in this book by taking this Final Test. Answers are given on page 182, following the exam.

1. Plant tissues are classified as either simple or complex. Listed below are the five simple plant tissues. Match the number of the tissue with the appropriate descriptive statement. All of the blanks require more than one answer.

   ________ a. Provide mechanical support to plant parts.

   ________ b. Typically made up of living cells.

   ________ c. Because of location and cell wall composition, these tissues are impervious to water and prevent water loss by the plant.

   ________ d. Contain a waxy substance in their cell walls.

   ________ e. Generally capable of becoming meristematic.

   ________ f. Cells of these tissues contain chlorophyll and can manufacture food.

   1. collenchyma
   2. cork
   3. epidermis
   4. parenchyma
   5. sclerenchyma

2. Characterize the relationship between the following items by listing them in order of increasing complexity.

   (A) tissues, (B) cells, (C) organs, (D) protoplasm

   ____

3. Meristems are sites of active cell division in a growing plant. Botanists recognize three kinds of meristems—apical, intercalary, and lateral. In the diagram on the next page, identify the apical and lateral meristems.

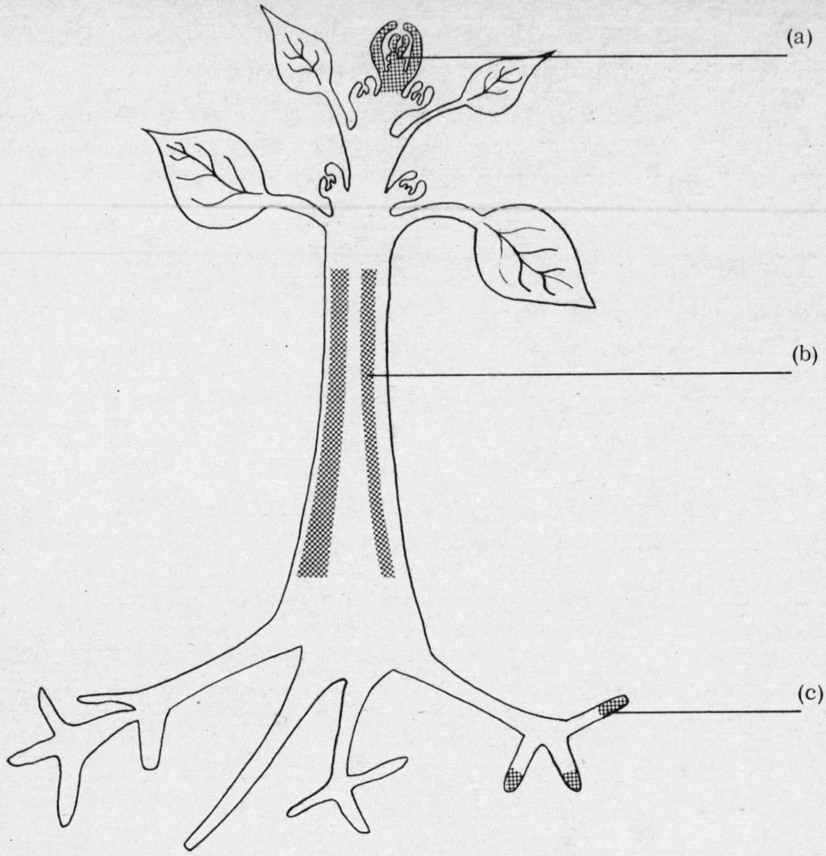

4. Numbered in reverse alphabetical order below are various kinds of cells found in complex plant tissues. Match the cell type with its corresponding description. Some of the blanks require more than one answer.

1. xylem parenchyma	5. sieve tube members
2. xylem fibers	6. phloem parenchyma
3. vessel members	7. phloem fibers
4. tracheids	8. companion cells

__________ a. By number, list all of the cell types that occur in the complex tissue that conducts food manufactured in the aerial plant parts to the roots.

__________ b. Bast fibers.

__________ c. Narrow, elongated, water-conducting cells with tapered, somewhat pointed end walls; dead at maturity.

__________ d. Dead cells with thick lignified secondary walls; end walls lacking at maturity.

__________ e. Characterized by having secondary walls that have scalariform, reticulate, or pitted thickening patterns.

(continued)

__________ f. List by number all of the cell types found in the water-conducting complex tissue of plants.

__________ g. Thin-walled parenchyma-like cells having pronounced nuclei; located adjacent to sieve tube members in flowering plants; precise function uncertain.

__________ h. Dead supportive cells with thick lignified cell walls; resemble tracheids.

5. Identify each of the following kinds of cells as being part of one of the five simple plant tissues—(1) collenchyma, (2) cork, (3) epidermis, (4) parenchyma, or (5) sclerenchyma.

________ a. Stone cells that give the flesh of the pear fruit its "gritty" texture.

________ b. Thin-walled cells in a potato tuber that store starch grains.

________ c. Cells covering the surface of a tomato leaf.

________ d. Cells containing suberin and produced by a lateral meristem.

________ e. Stomata and guard cells are a part of this tissue.

________ f. Found in the shell of the black walnut, making it quite hard.

________ g. Secretes the cuticle.

________ h. Tissue with unequally thickened primary cell walls that provide mechanical support.

6. Demonstrate your understanding of the origin of the parts of a flowering plant (for example, a bean plant) by matching the following numbered items with the lettered descriptions. Some of the blanks may require more than one answer.

________ a. Contain(s) stored food in bean embryo.

________ b. Give(s) rise to stem of bean seedling.

________ c. Source of root of bean seedling.

________ d. Source of xylem and phloem.

________ e. Gives rise to epidermis of seedling.

________ f. Cotyledons + epicotyl + hypocotyl + radicle = ____.

________ g. Forms parenchyma and collenchyma of seedling.

1. radicle
2. protoderm
3. procambium
4. hypocotyl
5. ground meristem
6. epicotyl
7. embryo
8. cotyledons

7. Shown below is a side view of a sectioned corn kernel. Identify the labeled parts by selecting their names from the numbered list.

1. cotyledon
2. embryo
3. epicotyl
4. hypocotyl
5. radicle

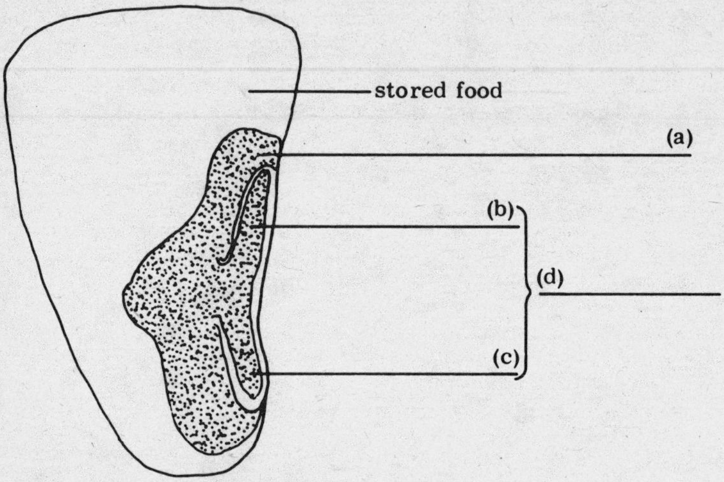

8. Refer to question 7 and answer each of the following questions.

   a. Part b will give rise to what part(s) of the corn seedling? __________

   ____

   b. Can you guess how part a functions? That is, what does it do?

   ____

   c. Part c will give rise to what part(s) of the corn seedling? __________

   ____

9. In the following drawing of the cross section of a root, identify the labeled items.

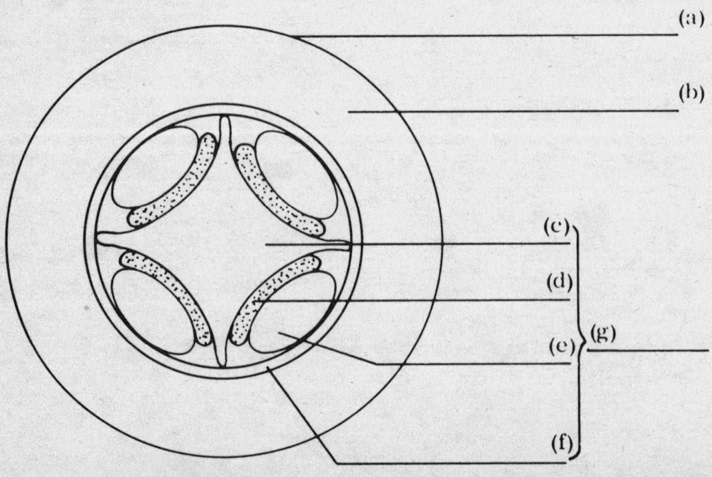

10. In the diagram in question 9, items c and e together are often spoken of as ____________________ tissue. Item f is important to the root for what reason? ____

____

11. Is the root cross section, shown in question 9, that of a monocot or a dicot? ____________________ Justify your answer. __________

____

12. When examining a root longitudinally, one can distinguish the various regions numbered below. Match these regions with the appropriate descriptions.

______	a. Region where cell number is rapidly increasing.	1. region of differentiation
______	b. Cells that protect the growing root tip.	2. region of elongation
______	c. Root hairs externally distinguish this part of the root.	3. meristematic region
______	d. Region where cell size is rapidly increasing.	4. rootcap
______	e. Region where epidermis, cortex, and stele first become evident.	

13. Roots may develop from the ____________________ of the embryo, or from the ____________________ of other roots, or adventitiously from such plant parts as ____________________.

14. Indicate whether each of the following statements is:

    1. true of stems only
    2. true of roots only
    3. true of both stems and roots

______ a. Branch endogenously.

______ b. Possess a cortex and epidermis.

______ c. Possess buds located at nodes.

______ d. Characteristically possess a well-defined endodermis and pericycle.

______ e. Cortex contains chlorenchyma.

______ f. Vascular tissue arranged in discrete bundles.

______ g. Have clearly defined stele.

15. Compare and contrast monocot and dicot stems by indicating whether the following statements are true of monocots (M), of dicots (D), or of both monocots and dicots (M & D).

_________ a. Vascular bundles characteristically lack a vascular cambium.

_________ b. Vascular bundles scattered throughout the stem.

_________ c. Chlorenchyma present in the stem.

_________ d. Stem has a well-defined pith.

_________ e. Stem has the potential to increase appreciably in diameter.

16. Corms, bulbs, rhizomes, and tubers are all examples of specialized stems that are often confused with roots.

   a. What characteristic do all of these specialized stems have which results in their being erroneously considered as roots? ____________

   ____

   b. List three external characteristics of these specialized stems that would clearly identify them as stems. ____________________________

   ____

   c. Study the three diagrams shown below and on the next page, and identify each as a corm, bulb, rhizome, or tuber.

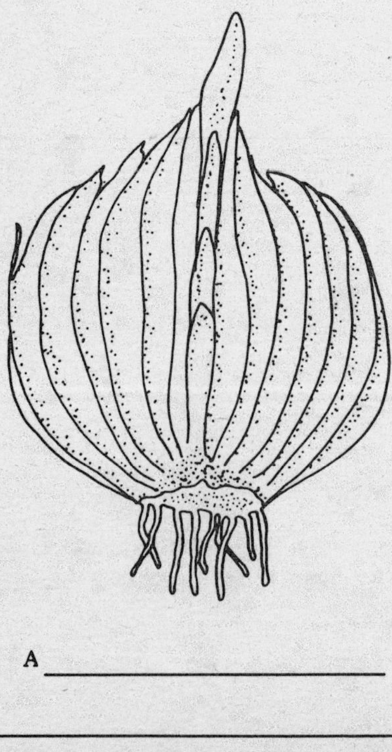

A ______________________________

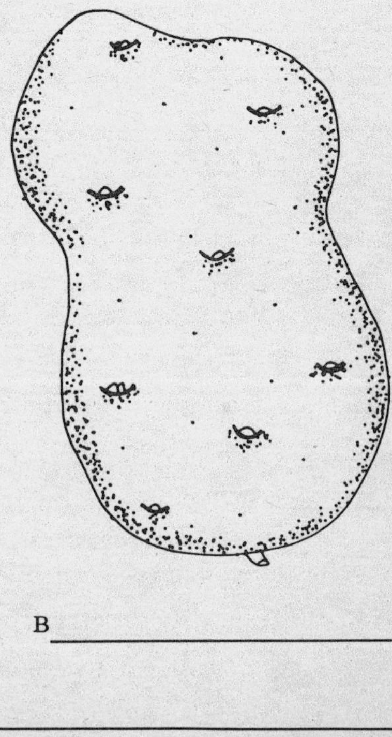

B ______________________________

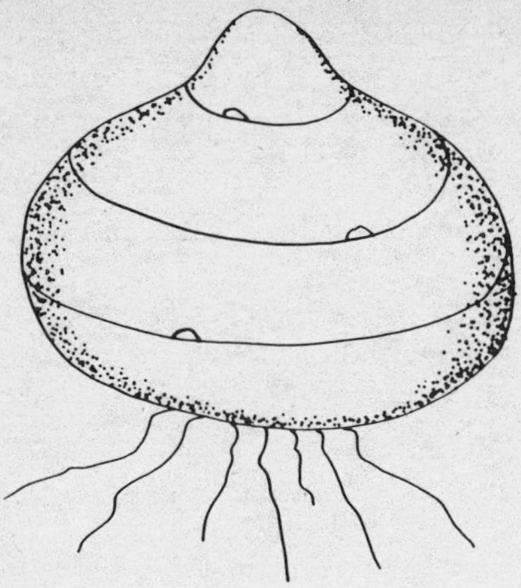

C ______________________

17. Shown below is a diagram of a typical woody stem. Identify each of the labeled parts in the diagram.

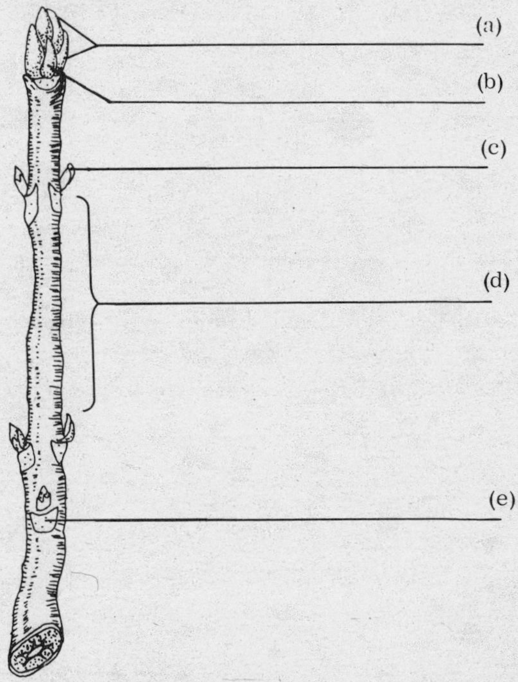

f. Item a in the above diagram will give rise to (leaves/stems/both stems and leaves) ______________________.

g. Item c in the above diagram will give rise to (stems/both stems and leaves/leaves only) ______________________.

18. Shown below is the cross section of a young dicot stem in which no secondary growth has occurred. Identify each of the labeled parts, selecting the parts from the following: vascular cambium, vascular bundles, cortex, epidermis, pith, phellogen, phellum, and pith rays.

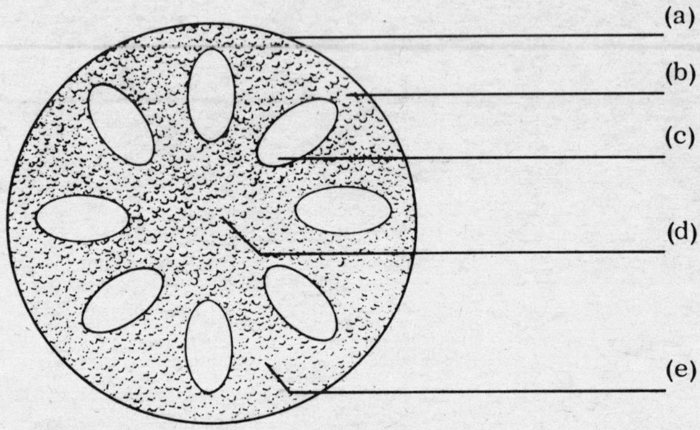

19. When a tree trunk is cut in various ways, we can study the cut surfaces and learn a great deal about the anatomy of wood. Basically three different sections may be made of the trunk—a cross section, a tangential section, and a radial section. Match the following words and phrases to indicate what may be learned from each kind of section.

________ a. Reveals annual rings as wavy lines.	1. cross section
________ b. Reveals annual rings as parallel lines.	2. radial section
________ c. Reveals concentric annual rings.	3. tangential section
________ d. Reveal vertical extent of wood rays.	

20. Shown below is a diagram of a young perennial dicot stem. Label the following items on this diagram: cortex, periderm, pith, secondary phloem, secondary xylem, and vascular cambium.

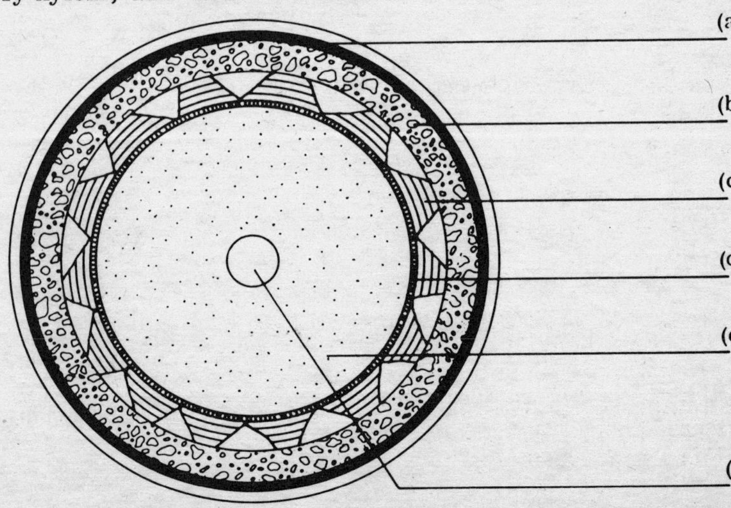

21. Now, identify each of the diagrams below as a cross, radial, or tangential section.

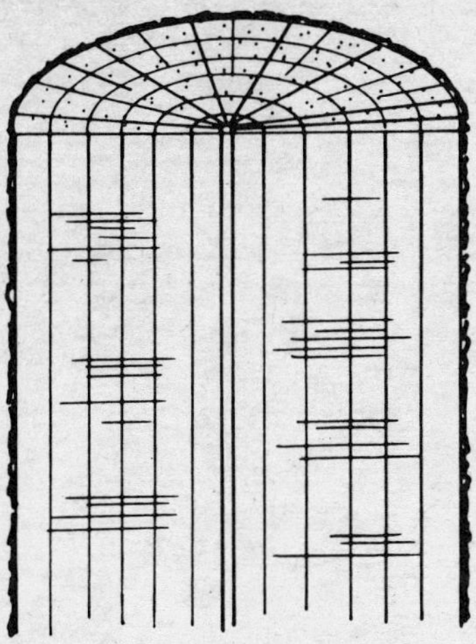

(a) ____________________

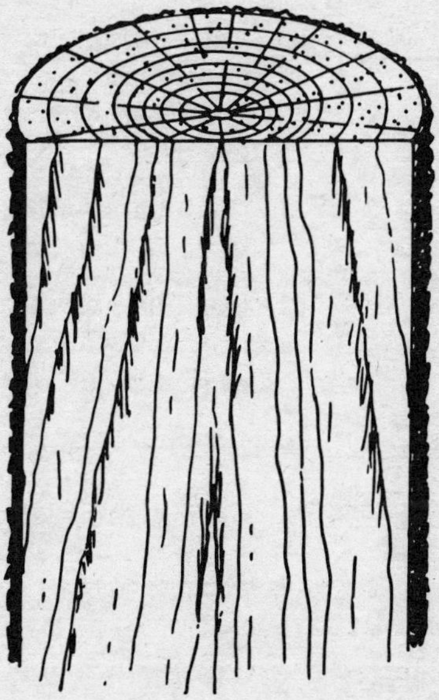

(b) ____________________

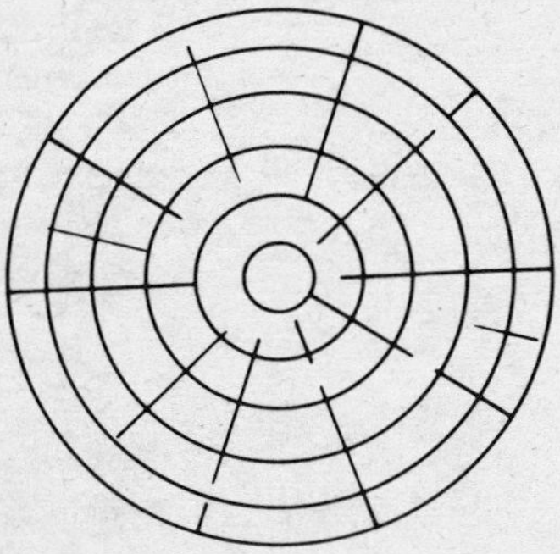

(c) ____________________

22. Listed below are a number of tissues or regions encountered in a typical dicot root or stem. Indicate whether each lies inside (I) or outside (O) the vascular cambium.

_______ a. cortex

_______ b. endodermis

_______ c. epidermis

_______ d. phellogen

_______ e. phellum

_______ f. primary phloem

_______ g. primary xylem

_______ h. secondary xylem

23. On the next page, complete the matching exercise to review some of the concepts about stem and root anatomy. Some of the blanks require more than one answer.

________	a.	Vascular cambium in both roots and stems gives rise to what tissues?	1. correct answer not given
________	b.	Vascular cambium of a perennial dicot stem arises in what two locations?	2. apical meristems
________	c.	Cork cambium gives rise to what two kinds of cells?	3. Casparian strips
________	d.	Cork cambium may arise in what two locations in a perennial dicot stem?	4. collateral vascular bundles
________	e.	Vascular cambium and cork cambium are both examples of ____.	5. cortex
			6. epidermis
			7. endodermis
			8. lateral meristems
			9. parenchyma cells of pith rays
			10. phelloderm
			11. phellogen
			12. phellum
			13. primary phloem
			14. primary tissues
			15. primary xylem
			16. secondary phloem
			17. secondary xylem

24. Check the correct answer(s). Pericycle in a perennial dicot root may develop into:

________ a. part of the cortex

________ b. branch roots

________ c. endodermis

________ d. vascular cambium

________ e. phellogen

25. Answer the questions below about the leaf diagram on the next page.

a. What name is applied to the broad flat part of the leaf, designated A in the diagram? ____________________

b. What name is given to part B? ____________________

c. Is the leaf shown a monocot or dicot leaf? ____________________

d. Give two reasons for your answer to question c. That is, what two characteristics of the diagrammed leaf permit you to decide whether it is a monocot or dicot leaf? ____________________

____

____

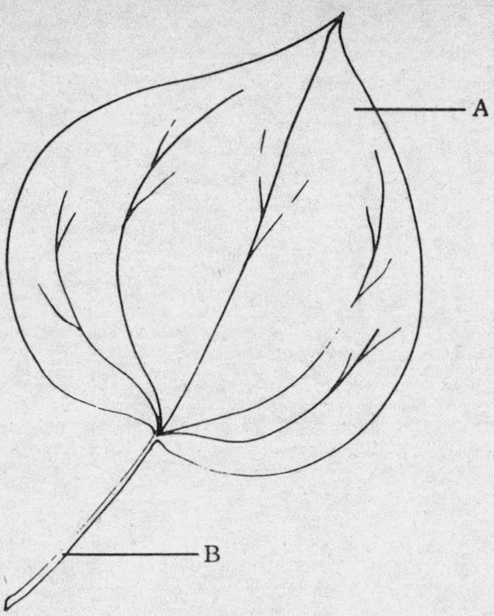

26. Identify each of the following parts of a leaf by giving the appropriate name for the part.

    a. Layer of cells covered by cuticle and containing stomata.

    ____

    b. Cylindrical cells near upper surface of leaf containing an abundant supply of chloroplasts. ____________________________

    c. Loosely arranged cells with many intercellular spaces; cells contain chloroplasts. ____________________________

    d. Cells described in b and c collectively constitute the middle of the leaf and are called ____________________________.

    e. Kind of cells or tissue constituting the bundle sheath of a vein in a leaf. ____________________________

27. Shown on the next page is a diagram of a flower with various parts identified by number. Answer the questions following the diagram by giving the number of the appropriate part.

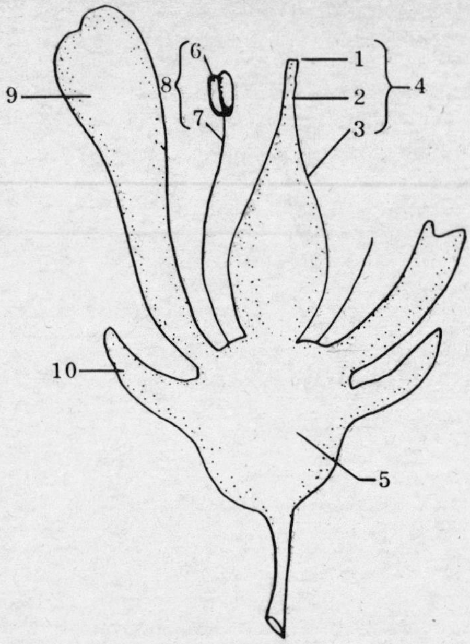

_______ a. In what specific part will seeds be produced?

_______ b. Of the four <u>major</u> flower parts shown, what two are essential for seed production?

_______ c. What specific part of the flower produces pollen grains?

_______ d. What specific part will mature into a fruit?

_______ e. What part is most leaflike?

What name is given to the following flower parts?

f. Part 6. ____________________

g. Part 9. ____________________

h. Part 10. ____________________

i. Part 4. ____________________

28. Shown below is a cross section of a typical wild type of tomato fruit. Answer the questions on the next page about this fruit.

a. How many carpels make up the pistil in this tomato? ____________

b. Although they are not shown in the diagram, how many dorsal bundles would you expect in this fruit? ____________

c. Assuming they have not fused, how many ventral bundles would you expect to be present in this fruit? ____________

d. What type of placentation does this fruit exhibit—axile, free-central, or parietal? ____________________________

e. The tomato is a fruit whose pericarp is entirely fleshy. The proper botanical name for this fruit would be (berry/drupe/pome/achene) ______________________________.

Answers

Compare your answers to the Final Test with the correct answers given below. If all of your answers are correct, congratulations! If you missed any questions, you may wish to review the chapter frames indicated in parentheses following the answer.

1. a. 1 (Chapter 1, frame 29) and 5 (Chapter 1, frame 31)
   b. 1 (Chapter 1, frame 29), 3 (Chapter 1, frame 37), and 4 (Chapter 1, frame 27)
   c. 2 (Chapter 1, frame 42) and 3 (Chapter 1, frame 37)
   d. 2 (Chapter 1, frame 42) and 3 (Chapter 1, frame 37)
   e. 1 (Chapter 1, frame 29) and 4 (Chapter 1, frame 26)
   f. 1 (Chapter 1, frame 29) and 4 (Chapter 1, frame 26)

2. D—B—A—C (Chapter 1, frames 12 and 13)

3. a. apical; b. lateral; c. apical (Chapter 1, frame 19)

4. a. 5, 6, 7, and 8 (Chapter 2, frame 23)
   b. 7 (Chapter 2, frame 22)
   c. 4 (Chapter 2, frame 4)
   d. 3 (Chapter 2, frame 5)
   e. 3 and 4 (Chapter 2, frame 8)
   f. 1, 2, 3, and 4 (Chapter 2, frame 14)
   g. 8 (Chapter 2, frame 21)
   h. 2 (Chapter 2, frame 12)

5. a. 5 (Chapter 1, frame 35)
   b. 4 (Chapter 1, frame 25)
   c. 3 (Chapter 1, frame 37)
   d. 2 (Chapter 1, frames 41 and 42)
   e. 3 (Chapter 1, frames 38 and 39)
   f. 5 (Chapter 1, frame 35)
   g. 3 (Chapter 1, frame 37)
   h. 1 (Chapter 1, frame 30)

6. a. 8 (Chapter 3, frame 9)
   b. 4 and 6 (Chapter 3, frame 12)
   c. 1 (Chapter 3, frame 10)
   d. 3 (Chapter 3, frame 4)
   e. 2 (Chapter 3, frame 4)
   f. 7 (Chapter 3, frame 5)
   g. 5 (Chapter 3, frame 4)

7. a. 1; b. 3; c. 5; d. 2 (Chapter 3, frames 6 and 15)

8. a. stem and leaves (Chapter 3, frame 15)
   b. absorbs food stored in the seed (Chapter 3, frame 15)
   c. root (Chapter 3, frame 10)

9. a. epidermis (Chapter 4, frame 14)
   b. cortex (Chapter 4, frame 14)

c. xylem (Chapter 4, frame 24)
d. procambial cells (Chapter 4, frame 26)
e. phloem (Chapter 4, frame 24)
f. pericycle (Chapter 4, frame 24)
g. stele (Chapter 4, frame 14)

10. conductive or vascular tissue (Chapter 2, frame 2); Item f, the pericycle, may become meristematic and give rise to branch roots. (Chapter 4, frame 33)

11. In all likelihood it is a cross section of a dicot root, since the central part of the stem is occupied by a stele consisting of xylem and phloem. Typical monocot roots have a central pith. (Chapter 4, frames 24 and 30)

12. a. 3 (Chapter 4, frame 9)
b. 4 (Chapter 4, frame 9)
c. 1 (Chapter 4, frame 13)
d. 2 (Chapter 4, frame 11)
e. 1 (Chapter 4, frame 14)

13. radicle (Chapter 4, frame 1); pericycle (Chapter 4, frame 33); stems and leaves (Chapter 4, frame 2)

14. a. 2 (Chapter 4, frame 32; Chapter 5, frame 4)
b. 3 (Chapter 4, frame 14; Chapter 5, frames 12 and 15)
c. 1 (Chapter 5, frame 6)
d. 2 (Chapter 5, frames 18 and 19)
e. 1 (Chapter 5, frame 16)
f. 1 (Chapter 4, frame 24; Chapter 5, frame 21)
g. 2 (Chapter 4, frame 14; Chapter 5, frame 19)

15. a. M (Chapter 5, frame 26)
b. M (Chapter 5, frame 25)
c. M & D (Chapter 5, frame 16)
d. D (Chapter 5, frames 20 and 25)
e. D (Chapter 5, frame 26)

16. a. All are underground stems. (Chapter 5, frames 30-36)
b. buds at nodes, attached leaves at nodes, and presence of nodes and internodes (any order) (Chapter 5, frames 31-36)
c. A: bulb (Chapter 5, frame 32); B: tuber (Chapter 5, frame 34); C: corm (Chapter 5, frame 31)

17. a. terminal bud (Chapter 5, frame 5)
b. bud scale (Chapter 5, frame 5)
c. lateral bud (Chapter 5, frame 5)
d. internode (Chapter 5, frame 6)
e. leaf scar (Chapter 5, frame 5)
f. both stems and leaves (Chapter 5, frame 8)
g. both stems and leaves (Chapter 5, frame 4)

18. a. epidermis (Chapter 5, frame 12)
b. cortex (Chapter 5, frame 15)

c. vascular bundle (Chapter 5, frame 21)
d. pith (Chapter 5, frame 20)
e. pith ray (Chapter 5, frame 21)

19. a. 3 (Chapter 6, frame 36)
b. 2 (Chapter 6, frame 37)
c. 1 (Chapter 6, frame 29)
d. 2 (Chapter 6, frame 37) and 3 (Chapter 6, frame 35)

20. a. periderm; b. cortex; c. secondary phloem; d. vascular cambium; e. secondary xylem; f. pith (Chapter 6, frame 20)

21. a. radial (Chapter 6, frame 37)
b. tangential (Chapter 6, frames 35 and 36)
c. cross (Chapter 6, frame 29)

22. a. O; b. O; c. O; d. O; e. O; f. O; g. I; h. I (Chapter 6, frames 18 and 25)

23. a. 16 and 17 (Chapter 6, frames 7 and 22)
b. 4 and 9 (Chapter 6, frames 20 and 21)
c. 10 and 12 (Chapter 6, frame 17)
d. 5 and 16 (Chapter 6, frame 24)
e. 8 (Chapter 6, frames 3 and 16)

24. b (Chapter 6, frame 9), d (Chapter 6, frame 10), and e (Chapter 6, frame 16)

25. a. blade (Chapter 7, frame 2)
b. petiole (Chapter 7, frame 2)
c. dicot (Chapter 7, frame 3)
d. The leaf has pinnate and net venation. The leaf is also attached to the stem by a petiole. These are both dicot characteristics. (Chapter 7, frames 2 and 3)

26. a. epidermis (Chapter 7, frame 9)
b. palisade cells (Chapter 7, frame 10)
c. spongy mesophyll (Chapter 7, frame 10)
d. mesophyll (Chapter 7, frame 10)
e. parenchyma (Chapter 7, frame 11)

27. a. 3 (Chapter 7, frame 30)
b. 4 and 8 (Chapter 7, frame 17)
c. 6 (Chapter 7, frame 20)
d. 3 (Chapter 7, frame 30)
e. 10 (Chapter 7, frame 17)
f. anther (Chapter 7, frame 20)
g. petal (Chapter 7, frame 17)
h. sepal (Chapter 7, frame 17)
i. pistil (Chapter 7, frame 17)

28. a. two (Chapter 7, frame 24)
b. two (Chapter 7, frame 24)

c. four (Chapter 7, frame 24)
d. axile (Chapter 7, frame 25)
e. berry (Chapter 7, frame 33)

Index

Pages indicating introductions, summaries, and Self-Tests are not included in this index.